GW00702995

Science: a second level course

S269 Earth and Life

Evolving Life and the Earth

Prepared for the Course Team by Peter Skelton, Bob Spicer and Allister Rees

The Open University

S269 Course Team

Course Team Chair	Peter Francis
Book Editor	Peter Skelton
Course Team	Angela Colling
	Nancy Dise
	Steve Drury
	Peter Francis
	Iain Gilmour
	Nigel Harris
	Allister Rees
	Peter Skelton
	Bob Spicer
	Charles Turner
	Chris Wilson
	Ian Wright
	John Wright
Course Managers	Kevin Church
	Annemarie Hedges
Secretaries	Anita Chhabra
	Janet Dryden
	Marilyn Leggett
	Jo Morris
	Rita Quill
	Denise Swann
Series Producer	David Jackson
Editors	Gerry Bearman
	Rebecca Graham
	Kate Richenburg
Graphic Design	Sue Dobson
	Ray Munns
	Pam Owen
	Rob Williams
Liaison Librarian	John Greenwood
Course Assessor	Professor W. G. Chaloner (FRS)

The Open University, Walton Hall, Milton Keynes MK7 6AA

First published 1997

Edited, designed and typeset by The Open University.

Printed in the United Kingdom by Jarrold Book Printing, 3 Fison Way, Thetford, Norfolk IP24 1HT

ISBN 0 7492 8185 5

This text forms part of an Open University Second Level Course. If you would like a copy of *Studying with The Open University*, please write to the Course Enquiries Data Service, PO Box 625, The Open University, Dane Road, Milton Keynes MK1 1TY. If you have not enrolled on the Course and would like to buy this or other Open University material, please write to Open University Educational Enterprises Ltd, 12 Cofferidge Close, Stony Stratford, Milton Keynes MK11 1BY, United Kingdom.

1.1

S269b4eli1.1

Contents

Preface

Frontispiece
Part of the reef around Mafia Island, Tanzania.

This book explores the history of interactions between life, as it has evolved, and the Earth, with particular emphasis on the impact of the evolutionary diversification of the eukaryotes.

Chapter 1 begins by reviewing the broad outlines of this history before going on to pinpoint the innovations in cell structure that conferred remarkable evolutionary exuberance on the eukaryotes. At this point we pause to consider how evolution by natural selection adapts organisms to the circumstances in which they live, and the implications of this process for the relationship between organism and environment. Then we delve into the distant past, looking at the evidence for the origin and early history of the eukaryotes, and their role in the emergence of complex marine ecosystems in the Proterozoic (2500 to 540 Ma ago). Chapter 2 follows on with a closer look at the nature and effects of the possible interactions between the late Proterozoic Earth and life.

Chapter 3 continues the exploration through the Phanerozoic (540 Ma ago to the present day), for which an abundant fossil record bears witness to a series of major periods of diversification, punctuated by mass extinctions, evidently linked with environmental crises. A step of major proportions investigated in Chapter 4 is the development of complex ecosystems on land, which added a new dimension to life's relationship with the Earth.

The interplay between changes in the Earth's geography and internal dynamics, and Phanerozoic life, is analysed in the next three chapters. General controls on climate, in particular, are outlined in Chapter 5. Two case studies follow, charting the Earth's passage through a cold, 'icehouse', period in the late Carboniferous and early Permian (Chapter 6), and a contrasting, 'greenhouse', period of widespread warmth in the Cretaceous (Chapter 7).

Chapter 8 briefly draws together the conclusions of the previous chapters to address the fundamental question concerning life's relationship with the Earth: benign partnership, or a chaotic system lurching from one temporary state of balance to another?

Chapter 1
The evolution of evolution

1.1 Introduction

> *... I look at the natural geological record, as a history of the world imperfectly kept, and written in a changing dialect; of this history we possess the last volume alone ...*
>
> *Darwin, C. R., 1859.* On the Origin of Species ..., *John Murray, London, p. 311.*

One aspect of Darwin's disappointment with the then known fossil record of evolutionary change was the apparent absence of Cryptozoic fossils. The only hint of such ancient life that he knew of, when he made the statement quoted above, was 'The presence of phosphatic nodules and bituminous matter' in them. Nevertheless, he firmly believed that there had to have been a long history of evolution prior to the seemingly abrupt first appearance of marine shelly fossils, as the latter already represented many different major groups of animals. His conviction has been vindicated, for we now have a fairly extensive, if partial, Cryptozoic fossil record, stretching back to Archean* times, though the story it tells might have come as something of a surprise to Darwin himself. It turns out that animals (multicellular organisms that ingest their food) do seem to have made a relatively late appearance, in the late Proterozoic, followed by the evolutionary explosion of shelly forms that announced the dawn of the Phanerozoic, which we will dwell on in Chapter 3. Chapters 1 and 2 concentrate on the preceding history that was such a mystery to Darwin – that of a strange world, which we now know to have been dominated until the latest stages by relatively simple organisms of limited diversity. Yet it was during this long earlier history that the fundamental pattern of feedbacks between life and the Earth, particularly the atmosphere, became established, and the precursors to the most conspicuous forms of Phanerozoic life – eukaryotes (whose cells contain nuclei) – evolved.

We shall begin this discussion of the evolution of life with a brief recap of how the Earth has changed over time (Section 1.2), followed by a look at the special features of eukaryote biology (Section 1.3), before focusing on the evolution of life during the Proterozoic from the first eukaryotes (Section 1.4) to the first animals (Section 1.5).

1.2 Former worlds

Earlier books in this Course have equipped you, in your imagination, to step into a time-machine and sample past conditions on the Earth.

Question 1.1

What conditions would you expect to find if you were to visit a lowland area on Earth (in mid-latitudes, say), (a) 4000 Ma ago, (b) 2000 Ma ago, (c) 500 Ma ago, (d) 100 Ma ago, and (e) 100 000 years ago? To help you think about your answer, decide which of the following items you would need to take with you: (i) breathing apparatus (with a supply of oxygen), (ii) suncream (factor 100), (iii) a thermally insulated self-contained capsule, (iv) a crabbing net, (v) weapons for fighting fierce monsters, (vi) sandwiches, and (vii) a good book.

---

* The Course style for the spelling of geological terms is explained in *Atmosphere, Earth and Life*.

The imaginary grand tour of Question 1.1 provides a reminder that certain aspects, such as mean surface temperature, have apparently stayed within modest limits for most of the Earth's history, although some have questioned this, postulating mean temperatures of over 60 °C for the Archean, at least. Other aspects – most notably atmospheric composition – have undergone radical changes. Life has apparently been present throughout virtually all the time considered, and as your earlier reading should have revealed, it has been implicated both in the regulation of global conditions and in their changes. Complex organisms individually visible to the naked eye are nevertheless of relatively recent vintage. Today, they present a kaleidoscopic web of interactions, both with each other and with their environments. The aims of this book are to explore how these relationships built up, and to analyse their consequences for the Earth and life as a whole.

For the first three-quarters of life's long history there is frustratingly little fossil evidence, despite its profound influence on the composition of the atmosphere, for example, but what there is shows surprisingly little change in outward form and diversity (Figure 1.1a). Yet this was a time of fundamental innovations in cell structure, giving rise to the first eukaryotes perhaps around 2100 Ma, or more, ago. Much later, a little over 1000 Ma ago, a prolific burst of eukaryote diversification ensued. The ecological relationships that subsequently emerged were to have revolutionary consequences for interactions between life and the Earth. Such meagre fossil evidence as there is, together with other kinds of clues, judiciously spiced with speculation, are the diet on which this and the next chapter must subsist for our exploration of these crucial formative events.

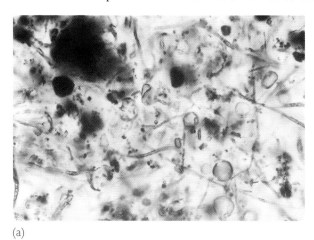

(a)

(b)

|—————————|
5cm

From around 540 Ma ago a proliferation of shelly animals began to litter the sea floor with their remains (Figure 1.1b), so inaugurating the rich fossil record of the Phanerozoic. The latter, closely studied now for over two centuries, has yielded a much more sharply focused picture of patterns in the history of life relative to the Cryptozoic record. Very uneven rates of turnover – of appearance and disappearance of groups – have been revealed, with mass extinctions episodically cutting across long phases of slower evolutionary change: one eminent paleontologist aptly commented, '… the history of any one part of the earth, like the life of a soldier, consists of long periods of boredom and short periods of terror' (Ager, 1993, p. 141). Indeed, this episodic pattern was widely acknowledged by geologists from the beginning of the 19th century, and was used by them as a basis for the relative geological time-scale that we still use today. Thus the most devastating mass extinction that we now know of was used to mark the close of

Figure 1.1
(a) Early Proterozoic fossil assemblage. Photomicrograph of prokaryotic cells in chert from the Gunflint Iron Formation, Ontario, Canada. Spheres are about 10 μm across. (b) An assemblage of shells of the trilobite *Redlichia* from the Lower Cambrian of Kangaroo Island, Australia.

the Paleozoic Era (discussed in Chapter 6), while a less pervasive but more abrupt mass extinction (including the dinosaurs) was taken to mark the end of the Mesozoic Era. Some of the geological periods within the eras were likewise terminated with mass extinctions, of varying effect, though not all were. Indeed, some mass extinctions fell within the recognized periods of the geological time-scale. Needless to say, some of these extinctions radically reset the agenda for subsequent evolution. The richness of the Phanerozoic fossil record, and the accompanying wealth of geological data, place constraints upon hypotheses seeking to explain such events and their aftermaths. They will be explored in greater detail in subsequent chapters, but for the present we must return to the roots of the eukaryote revolution. Before peering into the mists of Cryptozoic time, however, it is first worth reviewing what is special about living eukaryotes. An understanding of their basic biology is essential if we are to investigate their origins and make sense of the part they have played in the Earth–life system.

1.3 Empire of the eukaryotes

1.3.1 What is a eukaryote?

You were briefly introduced in *Origins of Earth and Life* to the distinction between eukaryotes and prokaryotes.

Question 1.2
What is the main difference in cell structure between these two major groups of organisms?

Figure 1.2
The arrangement of DNA in (a) a eukaryote cell, and (b) a prokaryote cell.

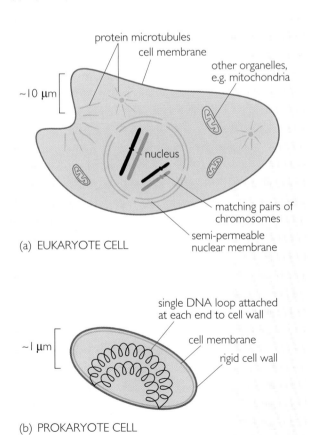

In addition to the basic difference noted in the answer to Question 1.2, the cells of eukaryotes possess orders of magnitude more DNA than do those of prokaryotes, packaged together with proteins to form a number of thread-like bodies in the nucleus called **chromosomes** (Figure 1.2a). Besides the nucleus, eukaryote cells also contain a host of discrete bodies called **organelles**, that carry out specific functions. Most prokaryote cells are surrounded by a rigid outer wall, and it is to the inner surface of this that their simple loop of DNA is anchored (Figure 1.2b). They also lack organelles.

Another feature of the eukaryote cell is an internal framework of protein rods (protein microfilaments and microtubules) that plays an important role in its internal organization.

For reasons that will be explained shortly, these differences in cell architecture, particularly in the arrangement of the DNA, have vastly expanded the scope of eukaryote evolution relative to that of the prokaryotes. The contrast is readily apparent from living examples (see Box 1.1, *overleaf*). All organisms that can be distinguished by the naked eye are eukaryotes. Most are multicellular, including plants, fungi and animals, though there are also many single-celled forms, reaching tens to hundreds of

Box 1.1 The kingdoms of organisms

Until well into the 20th century, it seemed suitable to divide organisms into only two broad kingdoms, plants and animals, the former mostly autotrophic and sessile (fixed), and the latter heterotrophic and mobile. As techniques of study improved, particularly with the advent of electron microscopes, it became clear that bacteria, which had traditionally been lumped with the plants, differed profoundly in structure from all other organisms. Hence the fundamental division between prokaryotes and eukaryotes was established (as outlined in *Origins of Earth and Life*). However, many eukaryotes still posed problems for the old plant/animal

dichotomy. Fungi, for a start, contradict their traditional 'plant' status by being heterotrophic, as well as showing various unique chemical and structural traits. Moreover, unicellular eukaryotes present a highly confusing picture. For example, some forms swim by means of a whip-like process called a flagellum, yet possess chloroplasts for photosynthesis, while other closely-related flagellate forms feed heterotrophically by ingesting prey. For a time, a convenient solution seemed to be to put all the unicellular eukaryotes in another kingdom, Protista, so leaving the multicellular forms to be divided between plants, animals and fungi. Further detailed studies of cell structure, and especially of differences in

DNA sequences, have caused even this neat compromise to break down, as further significant divisions became recognized, some cutting across the unicellular/ multicellular divide. At present, there is no universally agreed scheme, as evolutionary relationships are still debated. To avoid getting bogged down in detailed discussion, we will adopt the following pragmatic (if not strictly evolutionary) divisions of the eukaryotes in this book (differing in some details from the classification employed in *Atmosphere, Earth and Life*):

Kingdom Animalia – multicellular heterotrophic forms that ingest their prey.

Kingdom Fungi – largely multicellular (though with incomplete dividing walls) heterotrophic forms that

feed on organic molecules, which they either absorb directly or break down externally with enzymes and then absorb.

Kingdom Plantae – largely multicellular autotrophic forms, comprising land plants (all multicellular) and algae (both seaweeds, which are multicellular, and related unicellular forms).

Kingdom Protista – other unicellular forms, which may be either autotrophic or heterotrophic. As employed in this book, this grouping is still frankly a motley assembly of several only remotely related groups, some of which share relatively more recent common ancestors with certain of the other kingdoms.

micrometres (μm, or 10^{-6} m) in maximum dimension. Single-celled forms include amoebae, foraminifers, *Plasmodium* – the parasite that gives you malaria – and many kinds of unicellular algae. The vast majority of prokaryotes (e.g. bacteria), by contrast, are unicellular, though a few may group together to form multicellular filaments. Also, their cells are usually at least an order of magnitude smaller than eukaryote cells (rarely more than 10 μm in maximum dimension, and, in most forms, less than 1 μm). The innovations ushered in by the eukaryotes were therefore key events in ecological history.

1.3.2 Eukaryotes and ecology

Eukaryotes are clearly involved in a diversity of ecological interactions. Autotrophic forms (e.g. algae and land plants) photosynthesize to grow in a wide variety of habitats, while heterotrophic protists and animals consume them, and are consumed in turn by others: in your mind's eye, you could readily organize them in a hierarchy showing who eats whom – a **trophic pyramid** (Figure 1.3) – with the photosynthetic primary producers at the bottom and successive tiers of consumers stacked above them. Of the food available to each level, from lower levels, some is used up in respiration, and some is invested in growth and reproduction. A part, in turn, of the latter component then becomes the food supply for higher levels.

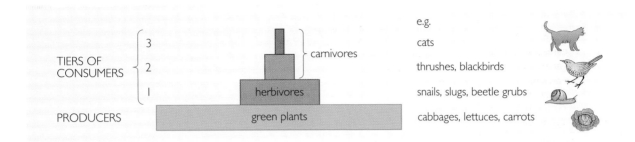

Figure 1.3
Simplified trophic pyramid. The width of each bar might represent the biomass, for example, or the amount of energy stored in tissues, at each trophic level.

■ This is a greatly simplified representation of feeding relationships among organisms. Can you think, however, of one crucial link that is missing from it?

■ What is missing is reference to the eventual chemical breakdown of any remaining organic materials – decomposition, in short.

Although some eukaryotes, especially fungi (Box 1.1), participate in this process, it is largely the province of bacteria, i.e. prokaryotes. They get everywhere where life reaches.

In addition to these agents of decomposition (and disease, where they invade living tissues), the prokaryotes of course also include major primary producers (i.e. autotrophic forms, such as cyanobacteria). But prokaryotes are essentially limited to obtaining energy in either of these two ways because of their rigid outer cell walls. Although dissolved molecules can pass through the cell wall (for heterotrophic feeding in decomposers or for synthesis in autotrophs), its rigidity prevents the cells from physically engulfing each other or any other solid particles. Hence they cannot ingest prey.

A world populated only by prokaryotes would lack the multi-tiered trophic pyramids discussed above. It would consist only of various kinds of producers and decomposers quietly flushing molecules in and out through their cell walls – rather a dull prospect! Yet such was the nature of ecosystems for the first half of the history of life on Earth.

1.3.3 Eukaryote architecture and evolution

Two attributes of eukaryote cells have been of paramount importance in the evolution of their diversity and complexity. First, their flexible cell membrane is supported internally by the molecular framework of protein rods, or microfilaments and microtubules (Figure 1.4). These can grow or shorten and so act rather like a system of extendible tent poles inside a tent. The eukaryote cell, unlike the prokaryote cell, can thus change shape and engulf external objects. This difference has permitted the addition of consumption by physical ingestion to the repertoire of feeding modes, and thus, ultimately, the build-up of multi-tiered trophic pyramids. The microtubules also play other crucial roles, for example in cell division, by controlling internal structure, as you will see below.

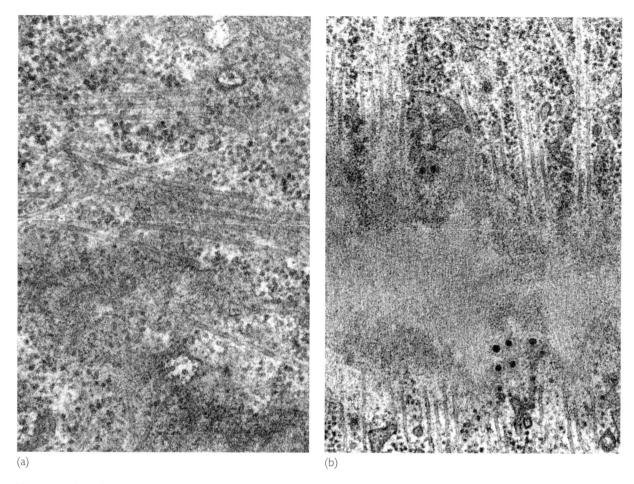

(a)   (b)

Figure 1.4

The internal framework of protein microtubules in eukaryote cells.
(a) Microtubules in a stem cell of Timothy grass (*Phleum pratense*) ($\times$ 66 000).
(b) Microtubules at the sidewalls of a pair of root cells in Thale cress (*Arabidopsis thaliana*) ($\times$ 52 000), a plant widely used in genetic studies.

The second attribute is the vastly greater library of DNA in eukaryotic cells which permits more specialized development. Multicellular forms, in particular, may contain a huge variety of cell types, making up tissues with widely different functions. Just consider those that make up your own body, for example. Specialized cells produce bones, skin and connective tissue for support and protection, while muscle cells allow movement, itself coordinated by nerve cells. A variety of secretory cells produce everything from hormones, controlling growth and behaviour, to digestive juices and mucus, while blood cells are involved in gas exchange and the immune system. Reproduction is effected by sex cells. And so the list goes on.

All these functions are underwritten by genes; nearly all cells possess the genetic information to make potentially any type of cell in the body. Development from a single initial cell means that the genes have not only to be reliably replicated as cells repeatedly divide, but their activity also regulated through interactions between genes, as different types of cell are produced. Such a complex set of instructions involves a lot more DNA than that found in the simple loop that suffices for a prokaryote. Whereas a typical bacterium may have some 4×10^6 base pairs in its DNA, that of a human, for example, has about 3.5×10^9 base pairs (though probably only up to a quarter of this actually codes for proteins, but that is another story). The management of so much DNA poses its own problems. For example, the total number of mutations is unavoidably increased, despite the existence of complex enzyme-controlled DNA repair mechanisms. For although the probability of mutation at any one DNA base site is extremely low (around one in a thousand million per replication), it is now compounded over thousands of millions of sites. Mutations are likely to arise at almost every cell division. Where these affect the functions of the proteins coded for, most will be of deleterious effect, and only a few, by chance, beneficial. Without some means of compensating for such errors, viable development of multicellular forms would be virtually prohibited.

One way the problem of harmful mutations is mitigated in eukaryotes involves a doubling up of chromosomes in the cell nucleus to yield what is termed a **diploid** complement of chromosomes: each chromosome thus has a 'homologous' partner, containing corresponding genes. If a gene in one set becomes corrupted by a mutation, then to some extent its function can be covered by the matching gene from the other set. Given the low probability of mutation of any one gene, the chances of both homologous genes being coincidentally affected is almost negligible. A twin-engined aeroplane provides an analogy: it may be able to continue flying even if one engine fails, whereas a single-engined craft will crash if its engine fails. Nevertheless, chromosome doubling by itself is an insufficient insurance against the risk of deleterious mutation *in the long run*.

Question 1.3
Can you think why this should this be so?

A further eukaryote innovation – **sexual reproduction** – gets around the problem discussed above by regularly mixing together genes from different individuals. Two steps are entailed in sexual reproduction (Figure 1.5). Fusion of special parental cells called **gametes** (e.g. sperm and egg cells), each containing only a single, or **haploid**, set of chromosomes, produces a new diploid cell (the fertilized egg). For the process to be repeated in subsequent generations, without continually doubling chromosome numbers, there clearly has to be a corresponding process to halve the numbers again between generations, i.e. to form haploid gametes. Normal cell division, known as **mitosis** (see Box 1.2), which is involved in body growth (or asexual reproduction in some forms), conserves chromosome numbers in the nucleus. But a special kind of cell division, **meiosis**, halves the number of chromosomes, whilst also exchanging segments of DNA between the matching chromosome pairs of the diploid parent (see Box 1.2). Together, meiosis to yield haploid gametes, and fusion of the latter to generate new diploid combinations, shuffle and distribute the genes of parents so as to deal a unique genetic hand to each of their offspring (except in the case of identical twins).

Figure 1.5
The cycle of sexual reproduction.

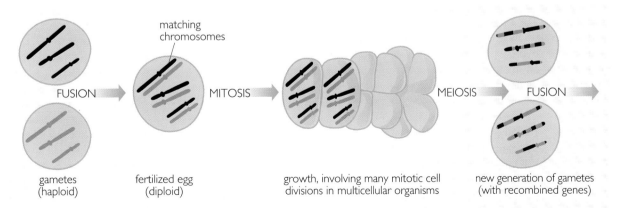

matching chromosomes

gametes (haploid)   FUSION   fertilized egg (diploid)   MITOSIS   growth, involving many mitotic cell divisions in multicellular organisms   MEIOSIS   FUSION   new generation of gametes (with recombined genes)

Box 1.2 Cell division in eukaryotes – mitosis and meiosis

An elaborate mechanism for cell division, called mitosis, has evolved in eukaryotes, that ensures the rapid duplication of DNA in the daughter cells with a high degree of accuracy. Prior to cell division, the DNA sequence of each chromosome (Figure 1.6a) replicates, the pairs of replicate strands for the time-being staying together (Figure 1.6b). A spindle-shaped structure of protein microtubules then forms within the cell, with the chromosomes aligned along its equator (Figure 1.6c). Meanwhile the nuclear membrane breaks down. The replicates of each chromosome now separate and are pulled by the microtubules to the opposite poles of the spindle (Figure 1.6d). There, each set of chromosomes becomes re-enclosed by a new nuclear membrane, while the spindle disintegrates (Figure 1.6e). Inward pinching of the outer cell membrane completes the process of mitotic cell division. This mode of cell division ensures that the huge amounts of replicated DNA are correctly divided between daughter cells, such that each receives an identical full set of chromosomes.

A second type of division produces haploid daughter cells, i.e. each with half the original number of chromosomes. This form of division is termed meiosis, and it forms the counterpart to fusion in the cycle of sexual reproduction

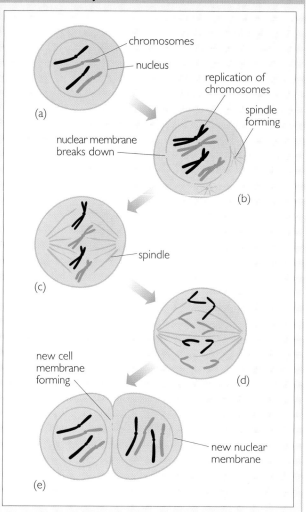

Figure 1.6
The process of mitosis. The chromosomes in (a) are different colours to indicate that they were originally from different individuals.

(Figure 1.5). Meiosis again employs spindles, though in two successive divisions. As before, the already replicated chromosomes (Figure 1.7a) become aligned along the equator of a spindle for the first division (Figure 1.7b). However, in contrast to mitosis, homologous chromosomes, derived from the fusion of parental gametes, are now also paired up alongside each other. A very curious process ensues. Each homologous pair (i.e. now four strands in all in each bundle, as each chromosome is already replicated) proceeds to twist along its length, in the manner of a towel being wrung out (Figure 1.7b). The replicated strands repeatedly snap, but their loose ends re-attach, though not necessarily to those from which they had just separated. Consequently, the four entwined strands randomly exchange

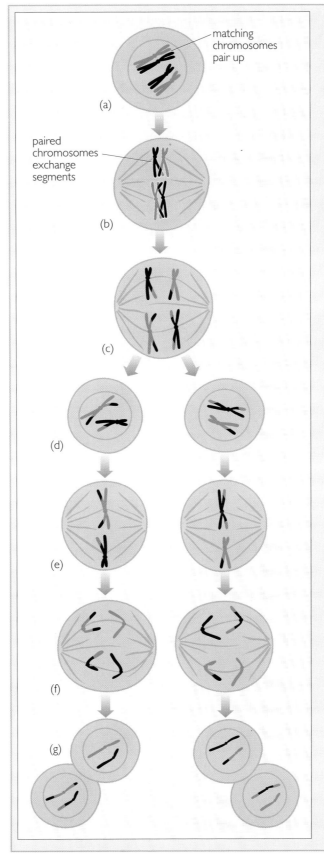

matching chromosomes pair up

(a)

paired chromosomes exchange segments

(b)

(c)

(d)

(e)

(f)

(g)

segments of their gene sequences with each other. Eventually, though, the chromosome pairs separate again (Figure 1.7c), and they now migrate to opposite poles of the spindle, to form two daughter cells (Figure 1.7d). The second division, of the daughter cells, is essentially mitotic in character (Figure 1.7d–g), separating the two strands of each chromosome (with their much shuffled sequences). A quartet of haploid daughter cells results, each cell containing only a single unpaired set of chromosomes (Figure 1.7g). Together, the crossing over of bits of the homologous chromosomes, and the independent assortment of the paired chromosomes into the daughter cells, bring about a thorough shuffling, or **recombination**, of the parental genes. No two daughter cells (e.g. sperm or egg cells) receive the same half share of genes. Harmful mutations can be effectively screened out by natural selection against those offspring that inherit them.

Figure 1.7
The process of meiosis.

The evolution of sexual reproduction in eukaryotes has had an enormous impact on the history of life, for it has altered the very rules by which evolution through natural selection proceeds. To a large extent, prokaryote evolution is a straight contest between clones (races of genetically identical individuals). Their reproduction by simple fission (involving straightforward replication of their DNA loop) produces virtually identical offspring, varying only through (rare) mutation. Many individuals must be produced to bring about a favourable combination of genes by coincidental mutation (Figure 1.8a). The small size and rapid reproduction of prokaryotes mean that the race can certainly be fast and furious, as clones with new advantageous mutations supplant others. And, in fact, things are not quite as simple as that, for some exchange of DNA between individuals can occur, providing an additional, if irregular source of genetic variation. However, sexual reproduction in eukaryotes provides a means for regularly varying the genetic complement that an individual passes on to its offspring, through recombination (Figure 1.8b). Potentially, all possible permutations of the genes available in a population can be assembled in different individuals and played off against one another in the Darwinian 'struggle for existence'. Those genes that generally combine to yield the best adapted individuals can be selectively assorted through the generations. A beneficial mutation can be rapidly promoted (i.e. its frequency in the population increased) by selection, unhampered by the cargo of the other genes present in the original individual in which it arose. The great leap in the efficacy of natural selection that this represents means that even large and complex multicellular organisms, with relatively slow rates of reproduction (compared with those of prokaryotes), can nevertheless evolve rapidly.

Figure 1.8
The effect of sexual reproduction on natural selection: (a) competition in asexually reproducing organisms is between clones, with variant offspring produced by rare mutations (represented by the dark symbols); (b) sexual reproduction furnishes each individual with its own unique permutation of genes, offering greater genetic variety for natural selection to act upon. The light and dark symbols represent alternative gene types. Only two genes on each chromosome are shown for the sake of simplicity. A high fitness combination of genes is one that through preferential survival and/or reproduction of individuals bearing it becomes relatively more frequent in subsequent generations. 'Fitness' is discussed in Section 1.3.4.

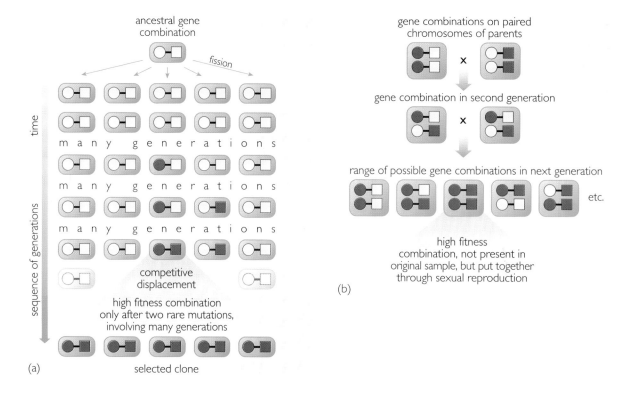

(a)

selected clone

(b)

high fitness combination, not present in original sample, but put together through sexual reproduction

The effect of sexual reproduction on the efficacy of natural selection, however, is strictly only a consequence of sexual reproduction having already become established in species. Although that effect has been of great importance in the long-term evolutionary history of the eukaryotes, it does not explain why sexual reproduction is maintained in populations, nor indeed how it might have arisen in the first place. There are several reasons why, at the individual level, asexual reproduction can offer advantages over sexual reproduction. For example, no energy (hence resources) need be wasted on finding a suitable mate, and achieving fertilization. Also, a female in a sexual population who 'cheated' and switched to asexual reproduction would endow each of her offspring with 100%, rather than only 50%, of her genes, and thus stand to pass on twice as many of her genes to subsequent generations. Moreover, proven favourable combinations of genes risk being broken up by sexual reproduction.

In view of these and other disadvantages, it is difficult to see how sexual reproduction could have gained a foothold and spread through populations at all: natural selection should surely suppress it because of the disadvantages of sexual individuals. Some ingenious hypotheses have been proposed by evolutionary biologists to counter such arguments. These refer to the possible advantages of sexual reproduction at the individual level. One model, for example, considers the effect of parasites. Rapidly reproducing parasites can soon adapt to a given combination of genes in their host organism. They can thus threaten the survival and reproductive success of all potential host individuals with that combination of genes. Asexual reproduction provides virtually no escape from the threat; the offspring are like a cursed race. Sexual reproduction, on the other hand, does offer an escape, by varying the genetic complements of offspring, such that some, at least, may prove robust, if not immune, to infection by the parasites. So there may well be advantages to sexual reproduction, offsetting the disadvantages, at the individual level, which could explain how it became established, and is maintained, in populations. Once there, it can provide the fortuitous gift of evolutionary flexibility, discussed above, to those species that have adopted it, and it is to this effect that the eukaryotes owe much of their evolutionary richness.

1.3.4 The rules of the new evolutionary game

It is worth following through the theoretical implications of the innovation of sexual reproduction for the evolution of adaptations in organisms. Analysing who or what benefits from adaptations will provide a useful perspective for our consideration of the relationship between the Earth and life.

The characteristic adaptedness of organisms is a product of evolution by natural selection, which may be summarized as follows. The reproductive potential of a population usually well exceeds the numbers that can be sustained, giving rise to Darwin's 'struggle for existence'. Individuals possessing heritable variations that promote their survival and reproduction in the prevailing circumstances, relative to other individuals, preferentially endow subsequent generations with their offspring and hence their genes. Given the genetic mixing of sexual reproduction, discussed above, genes associated with the favourable traits tend to increase in frequency in the population. Through time, the action of selection on the supply of variation hones such traits into adaptations of form and behaviour with recognizable functions in relation to survival and/or reproduction. The process is hierarchical, with individuals being selected, and genes associated with adaptive features consequentially sorted, generation by generation.

Who, or what, then, are the beneficiaries of the process – individuals, genes, or even whole populations, whose gene pools are affected? To try to answer this question, it is crucial to consider cause and effect. The efficient cause of the evolutionary process described above is natural selection itself, acting on individuals. There is an immediate pay-off, in the relative size of an individual's genetic legacy in the next generation. This is expressed as an individual's production of offspring, themselves surviving to be capable of reproduction, relative to other individuals – referred to as its **fitness**. The sorting of the genes, and any consequences for populations, however, are the incidental effects of such fitness differences. Hence the only logically necessary beneficiaries of the adaptations that evolve are the individuals possessing them, who are rewarded with higher average values of fitness relative to other individuals. (The 'selfish gene' concept of Richard Dawkins, set out in his book with that title in 1976, argues that it is the selected genes that get the ultimate benefit. Yet that is nevertheless still contingent on the successes of the well-adapted individuals carrying them.)

At first sight, two evolutionary phenomena may seem to contradict the conclusion above concerning the benefits of adaptations. The first is what biologists call 'altruism' within species (though with no implication of intentionality), and it involves one individual behaving in such a way as to benefit the reproductive success of another, though at a cost to itself. The second is the coevolution of different species to yield mutually beneficial features, such as that between flowers and their insect pollinators. To cut a long story short, closer analysis of both phenomena shows them to conform, albeit in subtle ways, with the conclusion above.

A classic example of altruism is the worker honey-bee, who feeds the queen and her offspring, and may die defending them, while failing to reproduce herself. How could such apparently selfless behaviour have evolved? The answer to this conundrum was provided in the 1960s by the British evolutionary biologist W. D. Hamilton. In most sexual organisms, the mean amount of genetic overlap between siblings (i.e. genes received in common from the parents) is 50%, because each individual gets a random 50% of each parent's genes (Box 1.2). Social insects such as honey-bees, however, have a rather curious pattern of genetic inheritance, because males are haploid, but females are diploid. Hence females (including the workers) receive all of their father's genetic complement, together with 50% of their mother's. Consequently, the majority (75% on average) of the genes received by each worker are shared with its siblings – the queen's other progeny, including new queen grubs. By helping the queen, and her daughter queens, who are especially adapted for copious reproduction, a worker ensures a far greater representation by proxy of her (shared) genes in subsequent generations than she might achieve by her own reproduction. So her effective genetic legacy, or 'inclusive fitness', which takes those shared genes into account, is increased by her apparently altruistic behaviour. Such traits are said to have arisen by a special form of natural selection termed 'kin selection'.

■   Why do you suppose the cells making up your own body do not compete with your sex cells to generate new individuals?

■   Since all the cells in your body are derived from a single fertilized egg, they all share the same genes. The sex cells therefore transmit the genes of all the other cells by proxy, leaving the other cells dedicated to their various specialized functions, as described earlier. The same principle can be applied to groups of individuals derived by asexual reproduction, and in which reproductive effort may be uneven, such as colonial coral polyps and vegetatively reproducing plants. These can all be considered as extreme examples of the kin selection principle.

Instances of coevolution, meanwhile, can readily be accounted for in terms of the fitness benefits separately accruing to the parties concerned. Each species in a coevolving pair is simply a regular part of the other's environment, and reciprocal adaptations arise accordingly: the most rewarding flowers are favoured by the insects, preferentially pollinated, and so selected for, while competition between the insects to obtain the rewards leads to selection for increasingly specialized features for so doing. Here, both participating kinds of organism happen to benefit from each other's adaptations. But such need not be the case, as is shown by another frequent outcome of coevolution, parasitism, where one participant suffers a loss of fitness because of the other (whose attentions it may not be able to avoid). The only way to make sense of such disparate outcomes of coevolution is to recognize that selection continues to operate discretely on the individuals concerned, with unpredictable effects at higher levels. Put another way, the insect and the flower cannot evolve as an entity, because each retains its own genetic identity, which is independently transmitted to its offspring, and subject to its own selective processes. The way in which genetic information is transmitted is thus the key to understanding the effects of natural selection and the adaptations that result.

Evolved adaptations are thus effectively of selfish, transient, benefit to the individual genetic entities concerned, and their rewards are strictly dependent on the prevailing circumstances. The adaptations provide no guarantee for the future welfare of the population (or even species) as a whole. A change in circumstances can precipitate extinction, as the testimony of the fossil record abundantly illustrates. Some adaptations may tend to prove deleterious to populations in the longer term, notwithstanding individual benefits in the shorter term. For example, in some environmental circumstances the acquisition of an asexual mode of reproduction by normally sexually reproducing organisms can lead to greater reproductive success, as explained earlier. However, those species that become fully asexual often appear to become extinct more rapidly than related sexually reproducing species. This is probably because of the associated decline in evolutionary flexibility of the asexual species. The individuals' short-term gain may eventually be at the longer-term expense of the population.

What is the implication of these arguments for the way in which living organisms interact with the Earth? Of great importance among the adaptations of organisms are systems of self-regulation, or **homeostasis** ('same standing'), that maintain stable conditions within the body in the face of a range of environmental perturbations. A familiar example of such a system is that which maintains your normal body temperature. In response to feeling cold, your body produces heat by various means (such as shivering), while it prevents over-heating by, for example, sweating. Thus you are equipped with an integrated system of sensors (temperature-sensitive nerves in this case) and compensatory devices that maintains a constant core temperature in the body. Such internal constancy is advantageous because enzymes, which are themselves adapted to function optimally in the body's normal state, may be highly sensitive to fluctuating conditions. It is tempting to try to draw a parallel between such self-regulation within organisms and the system of feedbacks that appear to regulate conditions on the Earth. Such a parallel has been explicitly proposed in the 'Gaia hypothesis' (Box 1.3).

Box 1.3 The relationship between life and the Earth according to the 'Gaia hypothesis'

The Gaia hypothesis, first formulated by James Lovelock in 1972, asserts that 'the climate and chemical composition of the Earth's surface environment is, and has been, regulated at a state tolerable for the biota' (Lovelock, 1989, p. 215). The Earth and its biota are regarded as having evolved together as a tightly coupled system, with self-regulation of important properties, such as climate and chemical composition, arising as emergent properties. The whole system is thus explicitly likened in this respect to a 'superorganism'. The kind of process Lovelock had in mind when proposing this hypothesis was the long-term maintenance of levels of molecular oxygen in the atmosphere sufficient to keep us alive, despite the short residence time there of this highly reactive gas (as discussed in *Atmosphere, Earth and Life*).

No theoretical explanation has been advanced as to why the feedbacks involved should necessarily serve to regulate conditions in the interests of the biota. Instead, a computer model to suggest how such a system might operate has been developed. Called 'Daisyworld', it envisages an idealized world similar to our own, receiving energy from a gradually warming Sun. The planet is imagined to have been seeded by two sorts of daisies, one black and one white. The essence of the model is that to start with, when solar heating is modest and the surface temperature is below the optimum for the daisies, the black daisies warm up more quickly than white ones and their growth is advantaged. As the black daisies spread, they reduce the planetary albedo, allowing more heat to be retained, and so help warm the atmosphere to an optimum level for the growth of daisies. Later, with increasing solar flux, optimum atmospheric temperatures may be surpassed in places, and it is the turn of the white daisies, whose greater reflectance helps keep them cool, to be at an advantage. As they now spread, displacing the black daisies, they increase the planetary albedo, and so serve to counteract the over-heating. Thus the planet and its biota together furnish a self-regulating system that, for a time at least, maintains optimum conditions for the daisies. Further refinements were later added to this model (with the addition, for example, of grey daisies, and a fauna of rabbits and foxes), none of which was found to disrupt the self-regulatory nature of the system.

To test whether Daisyworld is a valid model of our real world, we need to see if the feedbacks between life and the Earth have consistently operated over geological time in such a way as to maintain optimal conditions for life. We will return to this question in the final chapter of this book.

Question 1.4

Would you expect integrated homeostatic systems like those found in living organisms to evolve, by means of natural selection, above the level of the individual, operating, say, for the benefit of whole populations, ecosystems or even life as a whole?

It is important to distinguish between evolved homeostasis in organisms, of the kind explained above, and simple equilibration brought about by feedbacks. A mixture of ice and water will equilibrate at the freezing temperature of water, for example, because the bonds that hold water molecules together in the ice crystals provide a means of negative feedback. Any heat added to the system is taken up by some of the ice melting (a process that absorbs heat, as the bonds break). Loss of heat is compensated for by the heat released through the reverse process of ice formation. So long as both pure water and ice are present, the temperature stays constant. But the system has no checks and balances built into it to guarantee constancy in the face of other perturbations. The equilibrium temperature in this case can, for example, be altered simply by adding salt, which modifies the energy balance and lowers the freezing temperature. Without intervention, the system will not rid itself of the salt to retrieve the previous equilibrium temperature. By contrast, we do not expect our core body temperature to fluctuate (except temporarily, when ill), according to changes of diet, for example. Evolution by natural selection has furnished each of us with homeostatic devices that compensate for all the kinds of environmental vicissitudes that were faced by our ancestors (who did survive them and reproduce, as our own existence shows).

Equilibria in systems that do not reproduce in the manner of organisms, and have not therefore been honed by natural selection, lack such built-in safeguards to their stability.

Certain aspects of ecosystems, and indeed the Earth and life as a whole, may certainly equilibrate, because of the balancing effects of feedbacks (e.g. interactions between levels in the trophic pyramid) within them. But there is no reason to expect the essentially selfish adaptations of organisms to underwrite a given environmental equilibrium. Should conditions change, the winners in the 'struggle for existence' will simply be those that adapt to the new circumstances. Hence no support is forthcoming from natural selection theory, at least, for the claim of the Gaia hypothesis that the Earth and its life together regulate conditions *in the manner of a living organism*.

Nevertheless, the feedbacks between the Earth and its life have certainly yielded conditions quite unlike those that would have prevailed had the Earth been lifeless. The innovations introduced by the eukaryotes discussed above have allowed them in their turn to add their own significant twists to the tale, which we will now pursue.

1.4 The rise and rise of the eukaryotes

1.4.1 The prelude – stromatolite world

Let us return to the Earth of 2000 Ma ago that was conjured up in Question 1.1, and look more closely at the unappetizing greenish-grey mats along the coastline, or at least the record they left behind. The laminated calcareous structures they formed – stromatolites – dominate all but the last couple of hundred million years or so of the Proterozoic fossil record. They had already become quite widespread and diverse by early Proterozoic times. In rare cases, the fossil stromatolites contain tiny filamentous and spheroidal fossil remains (Figure 1.1a). But how representative are these microscopic fossils (**microfossils**) of the life-forms of that time? One of the best-known examples of such preservation, shown in Figure 1.1a, is the 2000 Ma old Gunflint Iron Formation that straddles the boundary between Canada and the USA around Lake Superior. The exquisite preservation of the fossil structures in this instance was due to their having been mineralized by chert (a form of silica, SiO_2, a more recent example of which is flint, found in chalk), precipitated from mineral-rich waters seeping through the mats. The mineralization must have taken place soon after burial, before significant decomposition of the organic materials could occur. The mineralization process was probably due to molecules of silicic acid (H_4SiO_4), formed by the earlier dissolution of silica, bonding with hydroxyl ions (OH^-) on the surface of the organic matter.

▣ What problem might such a style of preservation pose in connection with the question of whether or not these stromatolitic associations represent the main life-forms of the time?

▣ Somewhat unusual conditions must have prevailed in the original environment for the early mineralization to have occurred. Hence the associations themselves may have consisted of specially adapted organisms, unlike those living in other environments.

It is of course impossible to resolve this problem categorically, though the preservation of similar associations in Proterozoic rocks found elsewhere in North

America, Siberia and Western Australia suggests that they were at least widely distributed. The minute size of the fossil structures is consistent with a prokaryotic origin, the filamentous sheaths (usually 1–2 μm across) resembling those of living cyanobacteria, and the spheroidal structures (2–15 μm across), other bacteria. However, the filaments do not display the preferred orientations (alternations of horizontally and vertically oriented clusters) characteristic of mat-forming microbes in younger stromatolites, so interpretation of their relationships and ecology must be tentative. The common association of such stromatolites with banded iron formations, as in this example, is also of interest, implying the intermittent availability of both reduced iron (Fe^{2+}) and some molecular oxygen. The dissolved Fe^{2+}, as you read in *Atmosphere, Earth and Life*, was probably derived from hydrothermal vents on the ocean floor; it must then have dispersed widely through the water column, eventually precipitating in the oxidized state in the vicinity of the oxygen-producing cyanobacteria, in shallow inshore waters.

There are also some non-stromatolitic assemblages of microfossils in the Gunflint Iron Formation and elsewhere, preserved in silicified mudrocks. These include star-shaped forms interpreted as iron- and manganese-oxidizing bacteria, as well as simple, rather larger (6–31 μm across), spheroidal forms, of uncertain evolutionary relationships, which were probably planktonic.

A yet wider variety of stromatolites (ranging from flat-layered to domed, conical, columnar and even branched forms) is known from younger Proterozoic rocks, reaching a peak in those deposited around 1000 Ma ago. Microfossils also testify to a rich diversity of microbes then. The association of various fossils with differing sediments suggests some diversification among habitats. Many of the microfossils are quite similar to living cyanobacteria, or more rarely, to bacterial heterotrophs (decomposers). However, larger eukaryotic forms also contributed to the diversity of that time (Figure 1.9). Figure 1.10 shows a diagrammatic transect across a range of shelf-sea environments that can be interpreted from sedimentary rocks of mid- to late Proterozoic age, illustrating the variety of life-forms associated with them. The restricted inshore to supratidal zone (1) was dominated by tiny filamentous and spheroidal prokaryotes, producing stromatolites (like the earlier Gunflint examples). In open shallow subtidal settings (2), prokaryotic filaments were joined by larger (>50 μm), elaborately ornamented, eukaryotic cysts, called acritarchs (discussed below), which were probably planktonic. In somewhat deeper waters on the mid-shelf (3), less ornamented, spheroidal, acritarchs predominated. Over the outer shelf (4) and slope (5), there was a low diversity of small (<50 μm), simple, spheroidal plankton, with larger, probably eukaryotic, forms yielding entirely to clusters of smaller prokaryotic cells out in the basins. Because of the lack of preservation of ancient ocean floor, we have little idea of the life-forms there.

By mid-Proterozoic times, microbial ecosystems had already established most of the fundamental biogeochemical feedbacks that shape the modern world. Foremost among these was the evolution of a thriving plankton, which supplied a constant rain of dead organic material to the floor of deep marine basins. There, it could be incorporated in accumulating sediment, little disturbed by current activity, and not to see the light of day again until geological forces eventually returned the sedimentary rocks to the surface.

Question 1.5

Why was this burial of organic material of crucial importance in the evolution of the Earth's atmosphere?

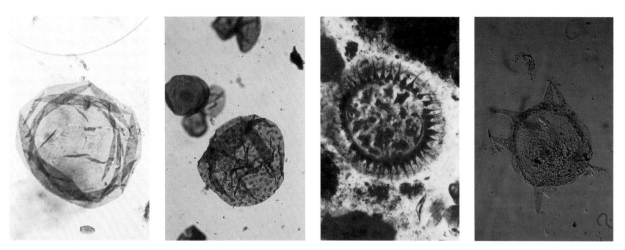

Figure 1.9
Fossils of early eukaryotes: a variety of cysts (resting stages), known as acritarchs, from rocks of late Proterozoic age. (Each box in the figure is about 300 μm high.)

Nor were the stromatolites, that so dominate the Proterozoic fossil record, without their own biogeochemical significance. Their widespread development within the photic zone, from open shelf to supratidal settings, trapped large quantities of calcium carbonate ($CaCO_3$), to form beds of limestone. This represented an important geological sink for carbon (in the carbonate), although not accompanied by the release of molecular oxygen, unlike photosynthesized organic matter. Nevertheless, the carbonate story was in time to be taken much further by the eukaryotes, as you will see later.

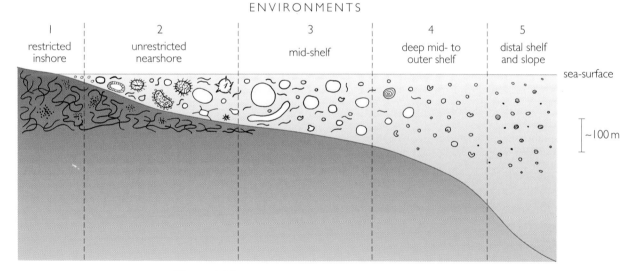

Figure 1.10
The reconstructed distribution of life-forms in a range of shelf to basin environments in mid- to late Proterozoic times. The five numbered environments and their inhabitants are described in the text.

1.4.2 Eukaryote beginnings

Fossil evidence of the earliest eukaryotes is extremely sparse. The first forms are likely to have been single-celled and devoid of any skeletal hard parts. Their potential for preservation would indeed have been slim. The discovery of exceptional fossils of such organisms therefore has a strong component of luck. Each year tends to throw up new records which push the inferred schedule for various evolutionary accomplishments further back in time. Even when found, however, the fossils may be difficult to interpret, as little critical structural information tends to be preserved. Criteria for recognizing eukaryotes are themselves prone to reassessment, as more is learnt about the characteristics of living forms and the processes involved in the preservation of the fossils themselves. Figure 1.11 summarizes the evidence available at the time of writing (1996): you may care to add to it any subsequent information that you encounter.

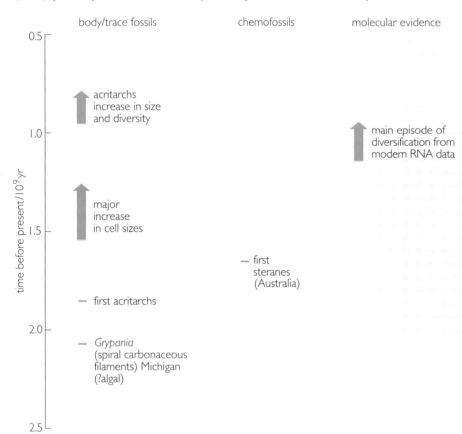

Figure 1.11
Evidence for the early evolution of eukaryotes.

What features should we look for to identify eukaryote fossils? Complex histories of chemical alteration of the original organic materials of the organisms, if preserved at all, effectively limits the analysis of body fossils to their structural features. The scale of resolution of these depends upon the circumstances of preservation. The finest preservation is found where organic materials have been directly coated, normally very soon (maybe only days or even hours) after death, by microscopic mineral crystals. The potential resolution of this style of preservation depends on the size of the latter, analogous to the grain in a photograph affecting the detail of the image. Rare examples of petrified soft tissues are known, from Phanerozoic deposits, in which even the shapes of cell nuclei can be discerned.

Usually, however, the preservation of structure is somewhat cruder. Where only impressions in the enclosing sedimentary rock remain, the grain size of the host rock itself limits the available resolution.

Question 1.6
From what you have learnt so far about the differences between prokaryotes and eukaryotes, which diagnostic features of the latter might you have some chance of detecting in fossils?

You might also have thought of the characteristic product of meiosis, a quartet of haploid cells (Box 1.2), and some Proterozoic fossil tetrahedral cell clusters have indeed been interpreted as such. However, because such clusters might equally well have been produced through successive fissions by prokaryotes that failed to separate, this criterion is regarded as unreliable.

In practice, none of the three possible criteria given in the answer to Question 1.6 has proved unproblematical in the search for the oldest eukaryote fossils. Dark spots within some Proterozoic fossil cells preserved in cherts have been interpreted by some researchers as representing organelles, but the development of similar features produced through cell collapse in rotting prokaryotes gives cause for doubt. Nor is evidence for cellular organization particularly pertinent in this context, as tissue differentiation only occurred much later in multicellular eukaryotes. Even cell size is not strictly diagnostic, as there is some overlap in the size ranges of prokaryote and eukaryote cells. However, significantly increased size in fossil cells is regarded as reflecting the spread of eukaryotes (see Section 1.3.1). In Proterozoic rocks deposited over the time interval from about 1600 Ma until some 1200 to 1400 Ma ago, the average size of fossil cells more than doubles. Marine shales dating from about 1400 Ma ago, moreover, yield the resistant organic cell coats, called acritarchs (Figure 1.9), which were mentioned in the previous section. Most of these probably represent the cysts (thick-walled resting stages) of eukaryotic algae that floated in the seawater.

Chemical fossils (chemofossils) push the origins of the eukaryotes even further back in time than the microfossil evidence cited above. Substances called steranes, formed from sterols (of which a familiar example is cholesterol), and known only from eukaryotic cell membranes, have been recorded from Australian petroleum deposits dated to at least 1700 Ma. The interpretation of yet older fossils remains inconclusive. Carbonaceous filaments, abundant in some formations dating from about 1300 Ma onwards, have also been recorded in rocks as old as 2100 Ma (*Grypania*, mentioned on Figure 1.11). These are commonly interpreted as multicellular eukaryotes, probably algae of some sort, but a prokaryotic identity for at least some cannot be ruled out. Recently reported examples from China, dating to about 1700 Ma ago, are considered to be eukaryotes.

All that can be reliably gleaned from the fossil record, then, is that the eukaryotes, as recognized on the limited criteria discussed above, had arisen at least by the mid-Proterozoic (by about 2100 Ma), if not some time earlier. How they evolved remains open to conjecture, although a fascinating hypothesis, first mooted some time ago, and which is consistent with a number of features seen in living eukaryotes, has been championed by the American biologist Lynn Margulis. Some eukaryote organelles bear striking resemblances to bacterial cells. **Mitochondria** (sing. mitochondrion), for example, which are responsible for the energy supply derived from aerobic (oxygen-consuming) respiration, contain small amounts of DNA, in a simple loop, like that in bacteria, although some of the DNA responsible for making them is housed in the cell nucleus. **Chloroplasts**, the organelles that effect photosynthesis in green plant cells, show a similar pattern. They can even undergo binary fission. From this and the evidence of other resemblances, Margulis suggested that these organelles were indeed once independent prokaryotes, which took up symbiotic residence inside ancestral eukaryotic host cells. This is known as the **endosymbiotic hypothesis** for the origin of eukaryotic organelles.

Question 1.7
Think back to what you read about natural selection and coevolution in Section 1.3.4. At what point would natural selection have ceased acting separately on the prokaryote precursors of these organelles, and their eukaryote hosts, such that they continued to evolve together as single integrated organisms?

Not all the organelles can be so readily explained thus. The nucleus, in particular, has its long chains of DNA in protein-bound chromosomes, as noted earlier, and shows other marked differences from bacteria. It probably originated from within the original cell structure of some distinct kind of prokaryotic ancestor (which thereby became the primordial eukaryote host cell type for the later acquisitions mentioned above). One effect of replacing a single DNA loop with multiple chromosomes is that the process of DNA replication during cell division can be carried out at the same time on the different chromosomes. It is thought that this may have been an adaptation for speeding up the business of replicating increasing amounts of DNA.

If at least some eukaryote organelles were thus secondarily acquired through endosymbiosis, the original ancestral host may have differed from other prokaryotes only in the possession of a nucleus. This primordial eukaryote line may well have extended back into the Archean, as a product of the early evolutionary diversification of the prokaryotes.

Question 1.8
Casting your mind back to the earlier discussion of the structural differences between eukaryote and prokaryote cells, what difference from the usual prokaryote condition would have been a prerequisite for endosymbiosis?

A living bacterium (one of the Archebacteria), *Thermoplasma*, which is adapted to extreme conditions of high temperature and acidity, lacks a rigid cell wall, and in this respect provides a plausible model for the eukaryote ancestral host cell. The ancestral form is thought likely to have been an anaerobically respiring heterotroph (Figure 1.12), that in due course was to adapt to rising oxygen levels through symbiosis with aerobic bacteria, capable of mopping up the oxygen, which eventually became mitochondria.

▪ What immediate and later advantages would have been gained by the host from such an acquisition?

▪ The aerobic bacteria would have helped remove the molecular oxygen, which would have been toxic to the anaerobic host. Later, the host could exploit the energy yield of its aerobic guests.

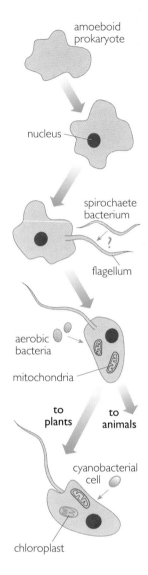

Figure 1.12
The sequence of possible acquisitions of organelles by eukaryotes according to a modern version of the endosymbiotic hypothesis. The first step, initiating the primordial eukaryote line, was the formation of a nucleus. Thereafter, some organelles at least may have been acquired through endosymbiosis of various prokaryotes, as shown.

Likewise, photosynthesis may have arisen in eukaryotes, through the acquisition of autotrophic symbionts that eventually became chloroplasts. These various acquisitions of organelle precursors were not necessarily unique events. The diversity of chloroplast types in different groups of algae imply independent acquisition in several different lines. In some cases, it even appears likely that the chloroplast precursor was itself a chloroplast-bearing eukaryote, because a nuclear relic is present within the chloroplast structure. The brown algae (e.g. kelp), so common along many shorelines, are one such group. There is as yet no evidence for mitochondria having been acquired more than once, but then again we know nothing of the mitochondria in the myriads of organisms that have become extinct.

The timing of this hypothesized evolutionary assembly of the variously kitted-out eukaryotes can only be guessed at, at present, as direct fossil evidence for the appearance of the organelles in question is lacking. Indeed the succession of endosymbiotic unions may well have extended over a long interval. But in the case of the spread of mitochondria, any time after the appearance of molecular oxygen in the atmosphere and oceans would seem plausible.

> **Question 1.9**
> On this basis, when, very approximately, might mitochondria-bearing eukaryotes first have begun to proliferate?

Such early evolutionary advances appear, however, to have had only a limited initial impact on the Earth's ecosystems – or at least those represented by the fossil record. For, as noted earlier, the largely prokaryote-dominated mats that yielded stromatolites continued to dominate shallow sea-floor communities for another 1000 Ma or so, until their rapid decline from around 850 Ma ago. The main evolutionary divergence of eukaryote stocks, including most of the various chloroplast-bearing groups, seems to have occurred only after a prolonged period of lurking in the wings.

1.4.3 Diversification of the eukaryotes

The fossil record of eukaryotes points to a marked proliferation of types in the late Proterozoic. Though known from older rocks, acritarchs (Figure 1.9), for example, become abundant and diverse in rocks dating from around 1000 Ma onwards. Andrew Knoll, from Harvard University in the USA, has led an intensive study of a sedimentary succession of marine origin, ranging in age from 850 Ma to 600 Ma, in the Arctic island of Spitsbergen. Fossils there record the presence of both prokaryotes and eukaryotes, from a variety of habitats. However, while the prokaryotes look very much like those living in corresponding environments today, and show little evidence for change, the eukaryote fossils present a different story. They reveal a marked diversification, with a relatively rapid turnover of species, as one moves up through the succession of sedimentary strata (**sedimentary succession**). Alongside single-celled forms, similar to living unicellular green algae and dinoflagellates, are also various multicellular algae (seaweeds). Conspicuous by its absence, though, is any fossil evidence for animal life, a point to which we will return shortly. Subsequent studies elsewhere appear to corroborate this story.

Another kind of evidence comes from comparing the DNA, or RNA, sequences of living eukaryotes (or the amino acid sequences of proteins derived from them). The idea, in outline, is that as evolutionary lineages diverge from a common ancestor, so too do their sequences, as a consequence of cumulative (non-deleterious) mutational change in each lineage. There are theoretical reasons, with some empirical support, for supposing that the average rate of such divergence remains approximately constant, at least for particular parts of the sequences. This proposal is known as the **molecular clock theory**, in reference to the supposedly clock-like gradual change of the sequences. In so far as it is correct (and there is plenty of debate about that), it implies that, for any three organisms, the pair that shows the smallest mutual divergence of sequences share the most recent common ancestor (Figure 1.13). On this basis, a hierarchy of evolutionary relationships can be built up for large numbers of living species, from which the shape of their evolutionary 'family tree', can be partially reconstructed.

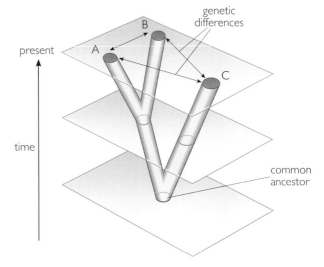

■   Why can such reconstructions of evolutionary trees be only partial?

■   They ignore extinct species, for which sequence data are not usually available. Though spectacularly newsworthy, fragments of fossilized DNA sequences, for example, are exceedingly rare, and analysis of them fraught with problems.

Absolute dating of some of the evolutionary branching points reconstructed from molecular data may be established from fossil evidence, where the record is sufficiently complete. Such calibration allows the average rate of change of the molecular clock in question to be estimated and the whole evolutionary tree accordingly set against an absolute time-scale. At the very least, the method provides a means of producing hypotheses concerning the pattern and chronology of evolutionary history that can be further tested against the direct fossil record.

Comparisons of certain RNA sequences from living organisms suggest that the majority of the major eukaryote groups separated from one another, in a flurry of evolutionary diversification from common ancestors, between about 1200 Ma and 1000 Ma ago. What sparked this burst of evolutionary activity is again open to conjecture. As with the origin of the eukaryotes, the fossil record bears witness to the aftermath, but is mute on how it came about. There is no geological evidence for any significant environmental change (e.g. of atmospheric composition or climate) around that time that might be considered to have precipitated this evolutionary explosion. A pertinent biological observation, though, is that those groups of living single-celled organisms that appear to represent the earliest branches of the eukaryote evolutionary tree are all effectively asexual (one such organism is the protist *Euglena*, a long-time favourite of school biology classes).

Question 1.10

What hypothesis to explain the main burst of eukaryote diversification might be deduced from this observation?

Though still only an hypothesis, the answer given to Question 1.10 highlights the important point that not all major evolutionary events need be explained by reference to external, environmental, triggers. Sometimes, internal modifications may fortuitously open up major new opportunities for further evolutionary change.

Figure 1.13
The relationship between the genetic differences between organisms and their times of divergence from common ancestors, according to the molecular clock theory. As the lineage leading to C split first from a common ancestor to A and B, its genetic sequence would have had more time to change independently than the lineages leading to A and B. Providing that the average rate of change in all lineages stayed the same, the sequences of A and B would remain more similar to each other than to that of C.

(a)   ⊢—⊢ 1 cm

(b)   ⊢—⊢ 1 cm

(c)   ⊢—⊢ 1 cm

(d)   ⊢—⊢ 1 cm

The evolution of sexual reproduction in eukaryotes is a likely example, indeed offering an explanation for what might be thought of as one of the biggest gear-changes in evolutionary history.

Whatever the circumstances, from that time onwards the overwhelming of the Earth's ecosystems by eukaryotes seems to have become almost unstoppable. Meanwhile, the diversity of stromatolites, which had reached a peak some 1000 Ma ago, began to decline, especially from about 850 Ma ago, although they remained widespread until Phanerozoic times, when they became largely restricted to stressful environments (e.g. highly salty bays or tidal flats) or cryptic habitats (such as cavities within reefs). Whether these changes reflect direct interactions between the prokaryotes and the eukaryotes, or whether stromatolite-building microbes declined in response to other environmental stresses, with various eukaryotes moving in on their vacated habitats, remains unclear. This is not to say that prokaryotes became unimportant. Far from it! Rather, they adapted to the eukaryote world in a variety of ways. Some entered into symbiotic relationships (including disease), and others adopted specialized roles, often in extreme conditions. Indeed, all life continues to rely upon the fundamental biochemical activity of prokaryotes for its continued existence.

1.5 The carnival of the animals

1.5.1 Ediacaran faunas

It was noted at the beginning of the last section that fossil evidence for animals in the first major episode of eukaryote diversification is lacking. Fossils widely interpreted as those of animals make a late but spectacular entry in rocks of the **Vendian Period**, representing the last 70 Ma or so (610–540 Ma ago) of the Proterozoic. From many parts of the world, assemblages of characteristic types of body fossils (Figure 1.14) have been recovered, which are collectively known as **Ediacaran faunas**, after the Ediacara Hills in South Australia, from where many such fossils were first described in detail.

With the exception of a few small simple forms in north-western Canada, dated to about 600 Ma ago, diverse Ediacaran assemblages range in age from at least 565 Ma ago until the end of the Vendian. A few distinctive types are known to have survived at least until mid-Cambrian times, especially in deeper-water habitats.

1.5.2 Interpreting the Ediacaran fossils

What were these organisms like? Perhaps their most striking feature is their size: some of the later forms reached up to a metre or so in length. Yet there is no evidence for their possessing any supporting skeletal hard parts. The fossils are found in sedimentary rocks interpreted as representing a variety of environments, ranging from deep marine areas to inshore, even tidal flat, settings. The fossils are preserved only as flattened impressions at the base of layers of originally sandy to silty sediment.

> **Question 1.11**
> Referring back to the discussion of fossilization at the start of Section 1.4.2, what level of resolution of anatomical detail would you expect from fossils preserved thus?

Figure 1.14
Examples of some Ediacaran fauna fossils. (a) *Dickinsonia*, an elongate pancake-shaped worm; (b) *Cyclomedusa*, a jellyfish; (c) *Tribrachidium*, a bun-shaped organism with three spiral tracts on its upper surface; (d) *Inkrylovia*, an elongate bag-like form with transverse partitions.

Paleontologists therefore have to interpret the Ediacaran fossils on the basis of their gross anatomical features. Needless to say, opinions have differed quite sharply. Earlier descriptions usually saw them being allocated to those phyla (major groups) still surviving today with which they seemed to have most in common (even if only superficially). Thus, various kinds of segmented annelid worms (similar to ragworms), primitive elongate arthropods (shrimp-like animals) and bun-like echinoderms (relatives of sea-urchins) were all recognized, as well as assorted cnidarians (jellyfish and colonial polyps). However, many of these identifications remained problematical. For example, in only one form, *Dickinsonia* (Figure 1.14a), have structures interpreted as guts been detected. Moreover, the intact condition of the fossils, even in sandstones that were deposited in shallow, agitated and presumably well-aerated, water, raises questions as to how they could have been preserved. Today, a combination of predation, scavenging and decay, not to mention disturbance by burrowing organisms, virtually rules out such a style of preservation of soft-bodied organisms in equivalent environments.

One explanation given for these exceptional assemblages is that large-scale predators, scavengers and deep burrowers simply did not exist at that time, making the Vendian a sort of privileged window of opportunity for preservation. Another proposal views the majority of Ediacaran organisms as a distinct eukaryote group, separate from the existing kingdoms, and of very simple construction. They are portrayed as having had a quilted mattress-like body, filled with fluid and enclosed in a tough cuticle (explaining their good preservation), and to have lacked guts: it is suggested that either they merely absorbed dissolved food molecules over their broad body surfaces, or that they may have contained autotrophic symbionts within their tissues, obviating the need for a digestive system. At least in the case of *Dickinsonia* (Figure 1.14a), however, this interpretation can be rejected. Apart from the possible gut structure, one specimen has been described as preserved with its segments torn across prior to burial. Yet the impression it created in the underlying sediment is just as deep as that made by intact specimens, showing that the body was firm and not, after all, fluid-filled. Another researcher has even suggested that the Ediacaran organisms may perhaps have been lichens (symbiotic associations of algae and fungi).

Most authorities, however, would regard the pendulum as having swung rather too far from the conventional view in such interpretations. Currently, the most widely accepted hypothesis is that most of the Ediacaran fossils do represent primitive animals, with the majority having a grade of organization equivalent to that of the cnidarians. Living cnidarians have a sac-like body wall directly surrounding a digestive cavity reached via a single orifice, as, for example, in a sea-anemone. The tough body wall consists of outer and inner cellular layers which are separated by a gelatinous layer. A few forms, in which segmentation and bilateral symmetry of the body seem to be well developed (Figure 1.14a), may perhaps be regarded as the primitive relatives of more advanced animals. Certain problematical forms, however, may yet turn out to be algae, rather than animals.

In addition to these originally soft-bodied forms, a few kinds of skeletal components are also known from the late Vendian, including some networks of sponge spicules (mineralized needles that provide a skeletal framework for many sponges).

The recognition of Vendian animals is consistent with molecular evidence that suggests an even earlier time of origin, using the molecular clock theory explained in Section 1.4.3. Figure 1.15 shows the estimated divergence of the amino acid sequences of the protein haemoglobin between various pairs of living animal species, plotted against the minimum possible ages for the original divergences from common ancestors, based on fossil evidence. The minimum age for divergence in each case is taken as that of the oldest fossil that can be attributed to one, but not to the other, of the evolutionary branches in question: the existence of such a fossil shows that the split from a common ancestor must already have taken place. Haemoglobin is the red pigment that transports respiratory gas molecules in our bloodstreams. Because there may have been successive amino acid substitutions at any given site on a haemoglobin molecule, values for the *cumulative* amounts of divergence may exceed 100%. These data were compiled by the Australian paleontologist, Bruce Runnegar, who went on to estimate the maximum value of haemoglobin divergence detectable among living animals to be about 190%.

Question 1.12

Using the molecular clock theory as a guide, extrapolate from the data shown in Figure 1.15 to deduce, approximately, the inferred latest time of origin of the animals, based on the maximum value of haemoglobin divergence given above. *Hint*: note that the origin of the animals can be assumed to have preceded the first divergence to have taken place within the group.

The value derived in Question 1.12 is, of course, only a minimum estimate, because of the possibility of yet more divergent forms having become extinct. On the other hand, the molecular clock may not itself have been regular, so the conclusion is by no means infallible.

The Ediacaran animals thus appear to represent a major evolutionary proliferation of large-bodied animals some time after the appearance of the first animals, which may date back to around 1000 Ma. In the next chapter, we shall consider how changes in the late Proterozoic world may have allowed their evolution.

Figure 1.15
Measures of the relative amounts of divergence in the amino acid sequences of haemoglobin between pairs of living animal species, plotted against the time by which divergence from a common ancestor must have occurred, based on fossil evidence.

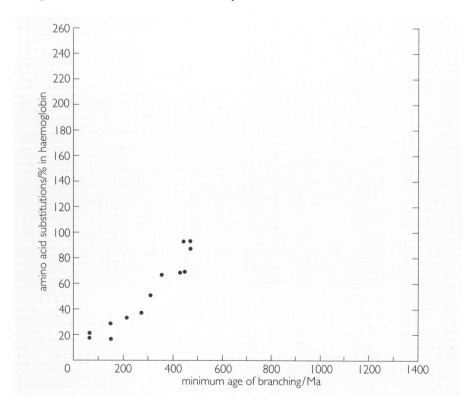

1.6   Summary of Chapter 1

1   While some conditions at the Earth's surface (e.g. mean temperature) may have remained within modest limits for most of its history, others (e.g. atmospheric composition) have undergone radical changes. Yet life has apparently been present almost throughout, though the main diversification of eukaryotes occurred only over the last 1000 Ma.

2   Eukaryote cell nuclei contain orders of magnitude more DNA, in chromosomes, than the simple loop attached to the inside of the rigid outer wall of prokaryote cells. The eukaryote cell also possesses an internal framework of protein rods, which can alter cell shape and control its internal structure. These differences have vastly expanded the relative scope of eukaryote evolution, to include multicellularity with cell differentiation, allowing them to build up multi-tiered trophic pyramids. Prokaryotes, meanwhile, have been confined to production and decomposition though under a wide range of conditions.

3   There is an increased probability of mutation associated with the increased amounts of DNA in eukaryotes. Compensation for this is provided by doubling of the chromosomes, together with the mixing of genes from different individuals through sexual reproduction.

4   Sexual reproduction has in turn increased the efficacy of natural selection, potentially allowing all possible permutations of the genes available in a population to be tested in the struggle for existence. Hence even complex multicellular organisms with slow rates of reproduction can evolve rapidly.

5   Adaptations that evolve through natural selection nevertheless remain of selfish benefit, in terms of fitness, to the individuals (or genetic entities) possessing them. They may influence higher levels, e.g. populations and ecosystems, but only by way of incidental effect, which may be good or bad. Hence, while natural selection can explain adaptations for homeostasis in individuals, the theory does not predict the emergence of analogous systems at higher levels, contrary to the claims of the 'Gaia hypothesis'.

6   Stromatolites dominate all but the last part of the Cryptozoic fossil record, reaching a peak around 1000 Ma ago and declining thereafter. Exceptionally preserved early examples contain microfossils of tiny filamentous and spheroidal prokaryotes. By mid- to late Proterozoic times, there was a clear differentiation of communities in different habitats, including both prokaryotes and simple eukaryotes. Planktonic forms supplied organic material to the deep sea floor.

7   Fossils of probable eukaryotes show that they had arisen at least by the mid-Proterozoic (around 2100 Ma ago). Organelles such as mitochondria and chloroplasts probably evolved from endosymbiotic prokaryotes that took up residence in the ancestral eukaryotic cells. The nucleus, by contrast, probably evolved within the original cell structure. The acquisition of mitochondria may have been in response to the first appearance of molecular oxygen.

8   Both fossils and molecular data point to an evolutionary explosion of eukaryotes commencing between about 1200 Ma and 1000 Ma ago. This may reflect the evolution of sexual reproduction: no particular environmental trigger has been implicated.

9   The last 70 Ma of the Proterozoic (the Vendian Period), was marked by the appearance of assemblages of large enigmatic fossils collectively referred to as Ediacaran faunas. They have been subject to a variety of alternative interpretations, but most are likely to have been primitive animals. Molecular data confirm the existence of animals from at least before the Vendian (from about 1000 Ma).

Chapter 2
Feedbacks between the late Proterozoic Earth and life

2.1 Introduction

What could have triggered the burst of evolutionary activity in the Vendian as described in Chapter 1? In contrast to the time of the earlier 'big bang' of eukaryotes (Section 1.4.3), there is frankly an *embarras de richesses* of evidence for environmental change in the last few hundred million years of the Proterozoic. Much uncertainty remains, however, as to the precise relative chronology of events, which obviously has to be sorted out if questions of cause and effect are to be resolved. As you might expect, there is no shortage of ideas.

We start (in Section 2.2.1) by looking at the inferred distribution of continents and oceans over the late Proterozoic which would have set the scene for such features as sea-level and biogeochemical fluxes, and hence climate (Section 2.2.2), as explained in *The Dynamic Earth*. We shall then consider the geographical and geochemical factors that are thought to have precipitated a series of major late Proterozoic ice ages (Sections 2.3.1 and 2.3.2). Finally, possible models for the feedbacks operating between the Earth and life are considered in Sections 2.3.3 and 2.3.4.

2.2 The setting

2.2.1 Late Proterozoic geography

The configuration of the late Proterozoic continents has to be reconstructed from evidence that is confined to the continents themselves.

- ■ Why is this so?

- ■ As a consequence of subduction, ocean floor is continuously lost through time, as old oceans close, to be replaced by newly opening ones elsewhere (*The Dynamic Earth*). Thus today's oceans are entirely floored by comparatively young (post-Triassic) crust.

Like the pieces of a gigantic jigsaw puzzle, ancient crustal regions (**cratons**) within the present continents may be reassembled to their earlier configurations. Clues for matching them up include fold belts that contain rocks of similar composition and age, corresponding sedimentary formations, and the shapes of their margins. Another clue is the 'remanent magnetism' preserved in certain rocks, due to the alignment of magnetic minerals with the prevailing magnetic field when they formed – they act like 'fossil compasses'. Measurement of the original angle of declination of the remanent magnetism allows one to estimate at what original latitude the rocks in question formed. The bearing of the remanent magnetism also gives an idea of how cratons have rotated on the globe.

Many issues still have to be resolved as far as the geography of the late Proterozoic is concerned. It is already clear, however, that a considerable rearrangement of tectonic plates was taking place over this time interval, involving both the rifting apart and the collision of cratonic masses. Around 1000 Ma ago, all the continents appear to have become clustered together to form one huge supercontinent, referred to as **Rodinia** (Figure 2.1), named from the Russian word *rodina*, meaning 'motherland'. However, subsequent rifting between the present western part of the North American craton ('Laurentia'), and a region comprising eastern Antarctica and Australia (along the line indicated in Figure 2.1), is now thought to have gone on to give birth to the Pacific Ocean. In effect, Laurentia seems to have burst out of the middle of Rodinia, setting in motion a protracted reassembly of continents through the late Proterozoic with a geologically rapid succession of continent–continent collisions. Some of these collisions brought together the South American, African and combined Indian/Australian/Antarctic landmasses to create a new supercontinent, **Gondwanaland**. Assembled by early Cambrian times, it was to persist for much of the succeeding Phanerozoic.

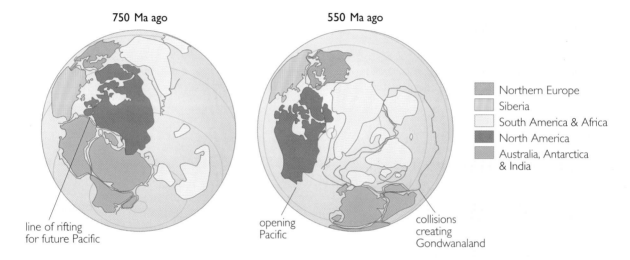

Figure 2.1
The break-up of Rodinia. Left: Rodinia, with incipient rifting. Right: with opening of the Pacific, today's Southern Hemisphere continents became assembled to form Gondwanaland. Note the differing orientations of the globe in the two reconstructions, with the South Pole near the bottom in the left image, and near the middle in the right.

2.2.2 Climate

What were conditions like on this mobile global framework? One remarkable feature is that sedimentary rocks of glacial origin appear to be represented on all the modern continents, although they make up only a subordinate part of their late Proterozoic sedimentary successions (see Box 2.1). At least three major ice ages have been proposed, separated by much longer, significantly warmer, periods. More than one glacial episode may have occurred within each ice age, and some of the glaciations must have been of extraordinary intensity, with some deposits suggesting the presence of ice at sea-level reaching even into equatorial paleolatitudes. The last of these, the **Varanger Ice Age** of earliest Vendian times (about 610–590 Ma ago), was perhaps the most extreme that the Earth has so far experienced.

Box 2.1 Deposits of the late Proterozoic ice ages

Diamictite is a descriptive term given to a sedimentary rock consisting of a poorly sorted mixture of gravel, sand and mud. Widespread diamictites of late Proterozoic age (Figure 2.2a) are interpreted to have been deposited by glaciers in a variety of settings, like the deposits of 'till' that can be seen beneath glaciers today (Figure 2.2b). Diamictites interpreted as glacial deposits are therefore also known as **tillites**. Some were deposited on land, but most were evidently deposited in sub-glacial lakes or in the sea, as melting glaciers, or vast ice-shelves and icebergs, dumped their loads of transported sediment. Typically included with the latter were boulders and pebbles that sank down into the surrounding sand and mud as 'dropstones' (Figure 2.2c). Some diamictites, however, formed not directly as till, but as mass flow deposits, collapsing from unstable slopes built up by the accumulation of sediments that may or may not have been of glacial origin.

A variety of other sedimentary rock types may be associated with the diamictites. Finely laminated silty mudrocks reflect seasonal discharges of turbid muddy water, while beds of sand grading up to mud were redeposited from subaqueous slumps of unstable sediment. Even stromatolites are known in some sedimentary successions, but unlike the more common marine examples generally associated with warm climates, they are interpreted to have formed in glacial lakes, like those that can be seen forming today in Antarctica.

The best known examples of such glacial deposits are those of the 'Varanger Ice Age' (about 610–590 Ma ago), which are particularly well represented in sites around the present North Atlantic, such as Scotland, Spitsbergen and Greenland. Most of the deposits in the western USA and Canada belong to the older 'Sturtian Ice Age' (about 750–725 Ma ago), and yet older deposits, of the 'Lower Congo Ice Age' (up to around 900 Ma ago) are known from Africa. All these ice ages are also represented in South Australia. Precise correlation of these deposits from area to area is problematical because of the lack of an accompanying fossil record. But the presence of distinct glacial units within the sedimentary successions (typically two for the Varanger Ice Age around the North Atlantic) points to there having been subsidiary glacial and interglacial episodes, as in our own Quaternary Ice Age.

(a)

(c)

(b)

Figure 2.2
(a) Late Proterozoic (Varangian) diamictites exposed in East Greenland. (b) Till sediments deposited by a modern glacier, Worthington Glacier in Alaska. (c) A Varangian marine glacial deposit in North-East Spitsbergen, with a dropstone at the centre. Outcrop is 7.25 cm high.

2.3 Why the ice ages?

2.3.1 Geographical factors

Several factors could have been involved in precipitating these drastic climatic changes. Consider again the movements of continental masses.

Question 2.1
Which aspect of the continental movements discussed earlier suggests one possible factor that could have contributed to these ice ages?

At present the relative dating of the ice ages and of late Proterozoic mountain-building episodes is simply not well enough constrained to test the hypothesis referred to in the answer to Question 2.1. We will return to other tectonic considerations in the next section.

2.3.2 Some geochemical considerations

Another line of evidence bearing on climate comes from the record of changes in certain stable isotope ratios that have been recorded by Andrew Knoll and his co-workers. Look at Figure 2.3, which summarizes the overall pattern of changes in the stable isotope ratios for strontium and carbon that have been observed in marine sedimentary rocks of late Proterozoic to early Cambrian age, together with other pertinent data.

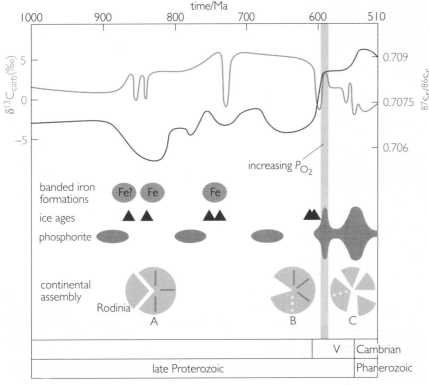

Figure 2.3
Strontium and carbon isotope curves for the late Proterozoic and early Cambrian, both derived from marine limestones. The strontium curve shows values for the ratio of the isotopes $^{87}Sr/^{86}Sr$. The $\delta^{13}C_{carb}$ curve shows deviations from a standard ratio of ^{13}C to ^{12}C (in ‰). Also shown are glaciations and times of deposition of banded iron formations, and of phosphorite deposits (discussed in Chapter 3), as well as cartoons of continental configurations. The bars in the cartoons represent rifting zones and the crosses are collision zones.

Consider first the curve for $\delta^{13}C_{carb}$. As this has been derived from marine limestones (i.e. the C in $CaCO_3$, known as C_{carb}), it can be taken to reflect seawater values of $\delta^{13}C$. For much of the time between about 850 Ma and 600 Ma ago, the $\delta^{13}C_{carb}$ curve shows relatively high values, though with marked negative excursions apparently associated with the ice ages.

Question 2.2
Cast your mind back to what you learnt in *Origins of Earth and Life* about the controls on carbon isotope ratios. Which carbon isotope must have been subject to preferential removal from seawater to account for the dominantly positive values of the $\delta^{13}C_{carb}$ curve, and how might this removal have been accomplished?

In broad terms, then, this period may well have been characterized by long-term net drawdown of CO_2 from the atmosphere, as it was incorporated into organic material that was being buried (rather than recycled). What might have caused this? Knoll and his colleagues argue that the break-up of Rodinia would have involved extensive rifting and formation of new ocean crust. This would have been accompanied by increased hydrothermal effusion of reduced chemicals (especially Fe^{2+}) into the oceans (*Atmosphere, Earth and Life*). By mopping up any molecular oxygen and so inhibiting the breakdown of organic material sinking to the ocean floor, these effusions would have augmented the rate of burial of the latter. Provided this burial of carbon was matched by the volcanic flux of CO_2 into the atmosphere, atmospheric levels of CO_2 could have remained more or less stable. Particularly intense episodes of carbon burial, however, could have led to significant depletion of atmospheric CO_2.

▓ From what you read about the effect of atmospheric CO_2 on surface temperatures earlier in the Course, what climatic consequences could be expected to have ensued from such intense episodes of CO_2 drawdown?

▓ As CO_2 is a greenhouse gas, atmospheric depletion should have led to climatic cooling. Such episodes could then have been the triggers for the glaciations of this period.

So far, so good, but why then should the glaciations themselves have been associated with sharp *negative* excursions of the $\delta^{13}C_{carb}$ curve? Consider the effects of the glaciations. By storing large amounts of water on land, as ice, the glaciations can be expected to have lowered sea-level. A likely consequence of this would have been the erosion and oxidative weathering of any organic-rich sediments contained within the newly-exposed marine strata. This, in turn, would have released isotopically light carbon (from the sediments) back into the ocean water, and could thus have caused the sharp temporary declines in marine $\delta^{13}C_{carb}$ values which are seen to follow hard on the heels of the glaciations themselves.

The other data shown on Figure 2.3 are at least consistent with this model for the inception of the glaciations. Look now at the strontium isotope curve. For the period prior to the time of the Varanger Ice Age, the $^{87}Sr/^{86}Sr$ ratio remained relatively low, mostly below 0.707. Such values suggest that the major source of strontium was from the mantle, via hydrothermal vents, instead of from weathering of continental rocks enriched in ^{87}Rb (the radiogenic isotope of rubidium), the decay of which yields ^{87}Sr (*The Dynamic Earth*).

Question 2.3
What are the likely implications of this conclusion for the hypothesis of increased continental weathering rates (with mountain uplift) causing climate cooling that was posed earlier in Section 2.3.1?

Whatever the case, there must have been considerable hydrothermal activity going on, either connected with faster sea-floor spreading or perhaps the development of a mantle superplume (*The Dynamic Earth*). Unfortunately, the scarce preservation of any oceanic crust from these times (in fold belts formed by continental collision) does not allow a choice between these two last options. Further evidence for increased hydrothermal emissions in general, however, is provided by the striking coincidence of some banded iron formations (BIFs) with at least the pre-Varanger glacial deposits (Figure 2.3).

Question 2.4
From what you read earlier in the Course, can you recall when, and under what circumstances, BIFs are believed to have been deposited?

2.3.3 Modelling the feedbacks

Together, the isotopic signatures, BIFs, and the inferred continental movements support a highly plausible model for the episodic but intense glaciations of this period (Figure 2.4). These episodes appear, in many respects, like brief returns of the BIF-forming conditions that had been so prevalent earlier in the Earth's history. Now, however, with a thriving plankton contributing massively to carbon burial, episodic glaciation also became a part of the story.

What has all this to do with the proliferation of animals? The period we have been considering (up to the Varanger Ice Age) is that which preceded their first visible burst of evolution – signalled by the majority of Ediacaran fossils. By inspection of the carbon and strontium isotope records, together with other data, through the Vendian, we can perhaps get some idea of the accompanying environmental changes.

▪ Refer again to the $\delta^{13}C_{carb}$ curve in Figure 2.3. Following the line of argument used earlier, what can you infer about rates of burial of organic material during the Vendian Period?

▪ At the start of the Vendian (610 Ma ago), the $\delta^{13}C_{carb}$ values of marine limestones were still relatively low, in association with the Varanger Ice Age (see Figure 2.3). Thereafter, however, marine $\delta^{13}C_{carb}$ values rose steeply again, suggesting a return to higher rates of burial of organic material (rich in ^{12}C).

Question 2.5
Now look at the other information pertaining to the Vendian Period on Figure 2.3. Again following earlier arguments, does it seem likely that the hydrothermal flux of reduced chemicals continued as before?

The Vendian world thus appears to have witnessed a significant change in the balance of fluxes in the oceans, with the resumed burial of organic material following the Varanger Ice Age no longer accompanied by such copious hydrothermal effusions of Fe^{2+}. If this were indeed the case, then important implications for atmospheric levels of oxygen may be surmised.

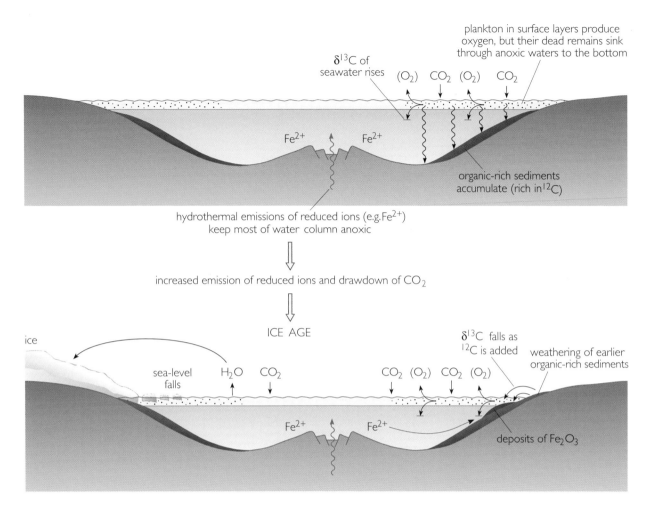

plankton in surface layers produce oxygen, but their dead remains sink through anoxic waters to the bottom

$\delta^{13}C$ of seawater rises (O_2) CO_2 (O_2) CO_2

Fe^{2+} Fe^{2+}

organic-rich sediments accumulate (rich in ^{12}C)

hydrothermal emissions of reduced ions (e.g. Fe^{2+}) keep most of water column anoxic

increased emission of reduced ions and drawdown of CO_2

ICE AGE

ice

sea-level falls H_2O CO_2 CO_2 (O_2) CO_2 (O_2)

$\delta^{13}C$ falls as ^{12}C is added — weathering of earlier organic-rich sediments

Fe^{2+} Fe^{2+}

deposits of Fe_2O_3

Figure 2.4
Knoll's model to explain the late Proterozoic ice ages.

Question 2.6
By reference to what you learnt about the links between the carbon cycle and the supply of molecular oxygen in *Atmosphere, Earth and Life*, how should the rapid burial of photosynthetically produced organic material have affected atmospheric oxygen levels?

The crucial step, as far as the Vendian is concerned, according to the hypothesis of Knoll and others, was the marked decline in hydrothermal activity. Prior to the Vendian, much of the surplus oxygen would have been used up in reaction with the reduced chemicals, especially Fe^{2+}, released into the oceans by the hydrothermal activity. So net delivery of molecular oxygen to the atmosphere would have been modest. The apparent decline in hydrothermal activity around the time of the Varanger Ice Age would have removed this constraint, allowing a rapid rise in atmospheric oxygen levels.

According to this model, the trigger for the evolutionary diversification of larger animals (which probably still relied largely on simple gas exchange, i.e. diffusion, to absorb oxygen) was the attainment of a threshold level of atmospheric oxygen (perhaps at least 1% by volume: *Atmosphere, Earth and Life*). Such a threshold could have been related to sheer size itself. If the shape of an organism remains constant, its surface area and volume increase at different rates with increasing size. For example, a doubling of linear dimensions (from L to $2L$) would yield four times the surface area for gas exchange (from L^2 to $4L^2$), but eight times the volume of tissue to be serviced by that surface area (from L^3 to $8L^3$). Hence as size increases (for a fixed shape), the efficacy of gas exchange declines. According to this model, then, increased atmospheric oxygen levels may have raised the upper size limit that primitive animals could attain. So the apparent flourishing of larger animals during the Vendian might signal their evolutionary emergence from the confines of small size enforced by the respiratory constraints of low oxygen levels. The simple disc and ring-shaped Ediacaran fossils found in sedimentary rocks pre-dating the end of the Varanger Ice Age, from north-western Canada (Section 1.5.1), are consistent with this analysis in so far as they attain only centimetre-scale diameters. The thin, pancake or flatworm shapes of the Ediacaran organisms could also have helped to increase the relative surface area available for gas exchange. Nevertheless, shape alone may not have been enough to satisfy the respiratory requirements of the largest forms. Bruce Runnegar has modelled those of *Dickinsonia* (Figure 1.14a) and concluded that some sort of circulatory system would also have been necessary to ensure adequate rates of gas exchange.

2.3.4 Other factors?

Even if correct, the oxygen threshold story is only one of a number of possible ways of linking global change at this time and the rise of animals. Another hypothesis, implicating the Varanger glaciations, has been mooted by the Russian paleontologist Michael Fedonkin. Ediacaran assemblages, he has noted, are largely associated with siliciclastic sediments (eroded sands and muds dominated by silicate minerals) of colder water origin, some even overlying glacial sequences. Regional disruption of the stromatolite-forming communities by the effects of the glaciations could thus have liberated shelf habitats for invasion by new communities of organisms, including the newly evolving stocks of Ediacaran animals. A problem with this idea, however, is that the main proliferation of Ediacaran faunas, in the late Vendian, followed some tens of million of years after the end of the Varanger Ice Age. So some continuing constraint would have been necessary in the post-glacial world to have held the stromatolite-forming communities at bay.

At present there is simply not enough information to choose conclusively between these, or various other, plausible models for the rise of the Ediacaran animals. However, in view of the evidence for major changes in conditions during the period leading up to their appearance, some kind of environmental stimulus for this particular evolutionary boom seems likely, in contrast to the earlier 'big bang' of eukaryotes. Far from being progressively stabilized in partnership with life, the record thus far shows that conditions at the Earth's surface have changed episodically, under the influence both of changes in the Earth itself and of feedbacks from life, and life has evolved in response.

2.4 Summary of Chapter 2

1   A supercontinent ('Rodinia') existed from about 1000 Ma ago until a series of rifting and collision events, which ensued over the late Proterozoic, led to its break-up and reorganization. During this period, a number of widespread glacial episodes were separated by longer, warmer, periods.

2   Relatively high $\delta^{13}C_{carb}$ values in marine sediments from around 850 Ma ago imply high rates of sedimentary burial of organic material. This may have been assisted by high rates of hydrothermal effusion of reduced chemicals,

especially Fe^{2+}, promoting oceanic anoxia. The glaciations may then have been caused by particularly intense episodes of drawdown of atmospheric CO_2. Negative excursions of the isotope ratio follow sharply, possibly because of weathering of the newly exposed organic-rich sediments as the sea-level fell.

3   Low $^{87}Sr/^{86}Sr$ ratios prior to the Vendian confirm the hydrothermal influence mentioned in (2), as also do the banded iron formations associated with the earlier ice ages. A sharp increase in the strontium isotope ratio during the last, Varanger, ice age suggests a lessening of hydrothermal influence. That would have meant less scavenging of the surplus oxygen generated when sedimentary burial of organic material resumed, allowing oxygen levels in the Vendian atmosphere to rise. A higher level of atmospheric oxygen has been mooted as a possible explanation for the evolutionary diversification then of larger animals, though other hypotheses have also been proposed.

4   The geological record shows that, far from being progressively stabilized in partnership with life, the Earth has changed episodically, under the influence both of changes in the Earth itself and of feedbacks from life, and life has evolved in response.

Chapter 3
Life's ups and downs in the Phanerozoic

3.1 Introduction

Reconstructing the pre-Phanerozoic history of life, as you saw in the last two chapters, depends upon rather patchily distributed, often problematical, fossils, and on somewhat speculative interpretations of other biological and geological data. The much more complete Phanerozoic record, by contrast, especially of marine shelly animals, allows quantitative estimates of the turnover in groups of organisms to be made. Thus it is possible, in particular, to identify the timing and scale of **evolutionary radiations** (phases of significant increase in numbers of species within groups of organisms), as well as **mass extinctions**, when abnormally large numbers of species became extinct together. Such information helps to narrow the search for cause and effect in the evolution of life.

The key to this improvement in the fossil record was the rapid evolutionary proliferation of animals with readily preservable skeletal hard parts or shells. The oldest examples are some tiny calcareous tubes, named *Cloudina* (Figure 3.1a) which are locally common in certain limestones dating from the late Vendian (from some 550 Ma ago). What is striking about the dawn of the Phanerozoic, though, is the sheer diversity of skeletalized organisms that evidently began to appear then. What sparked this explosion? What became of the earlier Ediacaran organisms? And what were the ecological and environmental consequences of this revolution? These are the main questions that we shall address in Sections 3.2 and 3.3. In Section 3.4, we explore the ensuing rich fossil record to see what pattern of relationships in time can be detected between evolutionary radiations, mass extinctions and major environmental changes. This allows us again to pick up the theme of the nature of the relationship between life and the Earth, seen from the context of the Phanerozoic record.

3.2 Ediacaran decline and the spread of consumerism

3.2.1 The Proterozoic–Phanerozoic transition

The transition from the Proterozoic to the Phanerozoic was a time of momentous change, both among organisms and in conditions at the Earth's surface. If we are to understand how these changes related to one another, it is first necessary to try to establish what actually took place, and when. Evidence for the chronology and nature of events sets the necessary constraints for disciplined speculation.

Let us start by considering the major faunal replacement that heralded the start of the Phanerozoic. There are three possible scenarios. First, the new Cambrian animals might have been the (distinctly modified) descendants of Ediacaran animals – a case of 'pseudoextinction', rather than true extinction (i.e. complete loss of issue), of the ancestral forms. Second, the newly arising forms, of unknown ancestry, could have driven the Ediacaran animals into extinction, either by direct competition or by predation upon them. Third, some kind of global perturbation of environments may have devastated the Ediacaran faunas, allowing the new arrivals to fill the ensuing ecological vacuum. Over the last few years, researchers have

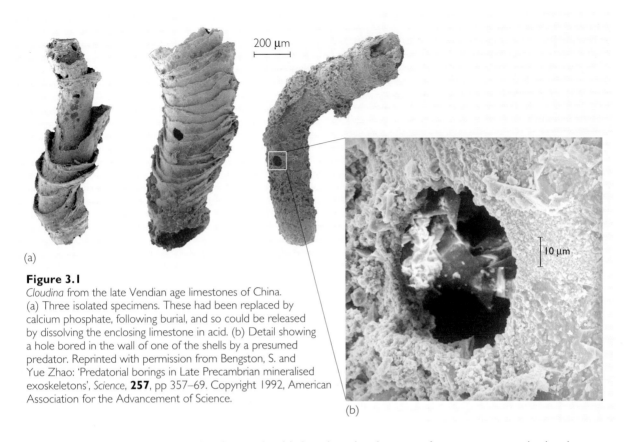

Figure 3.1
Cloudina from the late Vendian age limestones of China.
(a) Three isolated specimens. These had been replaced by
calcium phosphate, following burial, and so could be released
by dissolving the enclosing limestone in acid. (b) Detail showing
a hole bored in the wall of one of the shells by a presumed
predator. Reprinted with permission from Bengston, S. and
Yue Zhao: 'Predatorial borings in Late Precambrian mineralised
exoskeletons', *Science*, **257**, pp 357–69. Copyright 1992, American
Association for the Advancement of Science.

tended to favour the third explanation, because of an apparent gap in time between
the demise of the Ediacaran faunas and the Cambrian explosion. Recent work on
the Vendian to Cambrian sedimentary succession in Namibia, however, has called
into question that developing consensus, as explained in the next section.

3.2.2 Proterozoic Ragnarok?

The apocalyptic Norse myth of *Ragnarok* recounts the fall of the old order, in the
grip of strife amidst 'the winter of winters', followed by the repopulation of the
Earth by some hidden survivors. Could this be an apt metaphor for the close of the
Proterozoic?

Interpreting the geological record of the Vendian–Cambrian transition is beset by
two major problems. First, **correlation** of the sedimentary successions in different
parts of the world (i.e. recognizing beds of the same age from their fossil content or
other age-diagnostic features) is extremely difficult, precisely because of the
relative scarcity of fossils in the Vendian strata. Second, no one has yet discovered a
complete and unbroken sedimentary succession through this interval. Distinct
sedimentary gaps separate all the known sequences that contain Ediacaran faunas
from overlying Cambrian strata.

■ What is the most likely reason for such a widespread interruption of
sedimentary sequences?

■ Thinking back to what you read about the effects of sea-level change in *The
Dynamic Earth*, emergence of land, probably associated with a global fall in
sea-level, is the most probable explanation. This would have exposed older
deposits to erosion, prior to the resumption of deposition when sea-levels rose
again.

In many areas (e.g. Newfoundland, north-western Canada, eastern Europe, Siberia and China), the beds with diverse Ediacaran faunas are also separated from the Cambrian sequences by somewhat different sedimentary rocks. These intervening beds variously yield only an impoverished content of Ediacaran jellyfish-like fossils, various simple burrows, algal ribbons, and some small shelly fossils (like those shown in Figure 3.1a). In Siberia, stromatolites are also found at this level. Carbon isotope data give further clues.

Question 3.1

Study Figure 2.3 again: if you continue the line of argument followed in Section 2.3.2, what might the changing pattern in the carbon isotope curve over the last part of the Vendian and the start of the Cambrian imply about rates of burial of organic carbon?

The fluctuations imply at least considerable climatic instability at this time. The possibility even of a further glaciation (again perhaps provoked by the drawdown of CO_2) is suggested by the presence of glacial deposits in northern China that apparently post-date the Varanger Ice Age, but at most this seems only to have been a local development. By contrast, the Siberian stromatolitic limestones mentioned earlier probably imply a return to relatively warm conditions.

Question 3.2

Box 2.1 mentioned the possibility of stromatolite formation in glacial lakes. Study Figure 2.1 again and determine why these Siberian stromatolites are more likely to have formed in warm, rather than glacial, conditions.

Bitty though it is, such evidence hints at some climatic variability during the late Vendian, with marked fluctuations in the burial of organic matter. One likely circumstance leading to the extensive burial of organic matter over the interval would have been the stagnation, and associated anoxia, of deep waters confined within newly opening rift basins (see continental assembly diagram B in Figure 2.3). Organic-rich deposits of this age are indeed known from several parts of the world. The eventual disruption of such burial of organic material in the Cambrian could then be attributed to the greater circulation and overturn, hence oxygenation, of ocean water as these basins opened up and became connected (diagram C in Figure 2.3). Some other isotopic evidence (based on sulfur isotopes) points to mixing of previously anoxic bottom waters with oceanic surface waters at this time. Moreover, increased surface productivity of the plankton, fuelled by upwelling nutrients, is suggested by fairly widespread phosphate-rich marine shelf deposits (phosphorites) of latest Proterozoic to early Cambrian age (also shown in Figure 2.3).

From the kind of evidence outlined above, the story that seemed to be emerging over the last few years was of a pronounced gap in time, of many millions of years, between the decline of the Ediacaran faunas and the start of the Cambrian evolutionary explosion. But then, in 1995, American geologist John Grotzinger and colleagues published an important study of a fossiliferous sedimentary sequence of Vendian to Cambrian age in Namibia, known as the Nama Group. First they correlated the fluctuations of $\delta^{13}C_{carb}$ values in the Namibian sedimentary succession with those observed in sequences with Ediacaran fossils elsewhere, to establish the relative ages of different Ediacaran assemblages. In other words, they used isotopic 'signatures', instead of fossils, for correlation. Second, they determined absolute (radiometric) ages of volcanic ash bands in the Namibian sequence, so that they could assign absolute ages to the various assemblages. From this exercise they were able to provide a startling new chronology for the evolution of the Ediacaran faunas.

Importantly, they were able to demonstrate that Ediacaran fossils range through almost the entire Vendian. In particular, they concluded that the most diverse Ediacaran assemblages across the world all flourished within the last 6 Ma of the Vendian, persisting in Namibia virtually to the close of that period. The relative barrenness of terminal Vendian sequences elsewhere, in terms of Ediacaran fossils (as discussed above), was taken to reflect a failure of preservation, more than being evidence of a genuine extinction. Moreover, the late Vendian high diversity Ediacaran faunas were found to have been contemporaneous not only with *Cloudina*, but also with a small number of other small shelly forms. Hence the comfortable gap in time previously believed to have existed between the Ediacaran animals and those heralding the Cambrian explosion now seems to have disappeared, in Namibia at least. The possibility still remains, however, that Ediacaran animals were already in decline in the last few million years of the Vendian elsewhere, and that the youngest Namibian examples were in fact the last survivors; we cannot be sure at present, and the next few years will surely see intense study of late Vendian strata the world over in the search for an answer. What the Namibian study does do is to throw open again the possibility of any of the three scenarios for the Vendian to Cambrian faunal replacement outlined in Section 3.2.1.

3.2.3 The Cambrian explosion

In the past few decades a number of researchers have sought to explain the 'Cambrian evolutionary explosion' as a direct response to some change in the physico-chemical environment (e.g. to increased nutrient levels, oxygen availability, changes in oceanic chemistry affecting calcification, or expansion of shelf seas on the continental margins). Yet no clear link has been established with any of these factors. The influence of the environmental changes discussed above upon the Cambrian radiations might have been only indirect: by precipitating the demise of most of the Ediacaran organisms, they could effectively have liberated habitats for the surviving animals to exploit.

Another recently proposed model even implicates a form of positive feedback between the newly-evolving animals and their own radiation. It has been suggested that the tight packaging of organic detritus in their faecal pellets could have made it less accessible to heterotrophic (decomposer) bacteria in the water column and on the sea floor. This would have allowed the organic material in the pellets to have accumulated in sediments at an increased rate and so remain buried away there. The resulting increase in carbon burial and decline in the associated consumption of oxygen by the bacteria, could then have allowed increasing oxygenation of the water column. This in turn would have opened up still more habitat space for the evolutionary expansion of marine animals. We will reassess this hypothesis in a little while (Section 3.3.2) after taking a closer look at the life habits of the newcomers.

3.2.4 New body-plans and new ecological horizons

Apart from the extensive acquisition of skeletal hard parts, how did the newly radiating Cambrian organisms differ from their predecessors? And what ecological significance can be attached to the skeletal hard parts themselves? To answer the first question clearly requires some knowledge of the anatomy and life habits of the creatures concerned. You might suppose this to be a fairly tall order, if scattered shells and spicules were all we had to go on. But in fact, quite remarkably, the record is a lot better than that. Here and there, even soft tissues have become fossilized, in addition to skeletal elements, through chance combinations of factors. Such preservation may come about in a variety of ways, but appropriate conditions are often favoured by rapid burial, which protects the remains from any disturbance at the surface, either physical or biological, and from access to oxygenated water.

One such exceptional deposit, from the middle part of the Cambrian, is the Burgess Shale in the Rocky Mountains of British Columbia. Here, muddy sediments accumulating at the foot of a reefal edifice occasionally slumped into deeper anoxic waters, carrying off and burying some of the slope-dwelling animals. Decay of the soft tissues in this instance may have been inhibited by their intimate enclosure in certain clay minerals, which could have adsorbed destructive enzymes on the mineral surfaces and so arrested their chemical activity. The exquisite preservation (illustrated in Figures 3.3 and 3.4) and diversity of these often bizarre-looking creatures (Figure 3.2) provide an invaluable opportunity to compare the products of the Cambrian evolutionary explosion with their Ediacaran counterparts.

Figure 3.2
Reconstruction of some of the Burgess Shale fauna in life, viewed as in the beam of a time-traveller's bathyscaphe. The foreground is cut across to reveal some worms in their burrows; that on the right is *Ottoia*, also illustrated in Figure 3.3. Other forms, dwelling on the surface, are largely arthropods (with an external skeleton and paired jointed limbs), though velvet worms with fleshy 'limbs' (centre and right background), an obscure land-dwelling group today, as well as the multi-plated creeping worm, *Wiwaxia* (right, centre) and small hyolithids (right, foreground), distantly related to molluscs, are also present. In the background various sessile suspension-feeding animals (mainly sponges, but also primitive echinoderms at left background) project up from the sea floor.

Question 3.3

Compare the animals from the Burgess Shale, particularly the mobile forms shown in Figure 3.2, with the Ediacaran examples illustrated earlier in Figure 1.14. What broad differences can you detect between the two kinds of fauna, in terms of (a) overall body shape and symmetry, (b) differentiation of structure in different parts of the body, and (c) presence or absence of detectable limbs or other protuberances? (You don't need any prior technical knowledge to answer this question; just compare what you can see in the figures.)

You might readily imagine most of the Ediacaran organisms merely lying upon, or projecting up from, the sea floor. Perhaps a few might have crept across it (e.g. *Dickinsonia*), while jellyfish gently pulsed through the water above. There is barely any suggestion, however, of sensory, or food-trapping, organs being concentrated at a 'front' end. By and large it looks as if any resources upon which they relied came to them, rather than that they actively sought them out. Many of the Burgess Shale animals, by contrast, were clearly equipped to scuttle over, or burrow well into, the sediment, or even perhaps swim around with flexural motions of the body and/or limbs, in the pursuit of food. The differences in life habits appear to reflect a significant upgrading in body architecture.

The lack of evidence for guts in the Ediacaran fossils (with the possible exception of *Dickinsonia*) was already noted in Section 1.5.2. Among the Burgess Shale fossils, by contrast, are clear indications of discrete tubular guts (Figure 3.3: *Ottoia*), with openings (mouth and anus) at each end. The gut wall would originally have been separated from the outer body wall by a fluid-filled cavity. The muscular gut wall of comparable animals today allows the passage of food through the gut to be decoupled from the external motions of the animal (unlike, say, sea-anemones).

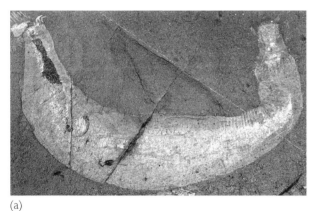
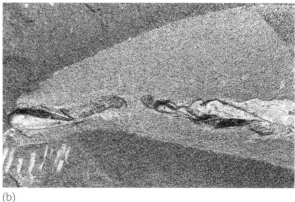

(a)   (b)

Figure 3.3
(a) *Ottoia prolifica*, a worm from the Burgess Shale (× 1.8). (b) Detail showing shells in the posterior part of its gut (× 3.5).

Such internal, fluid-filled body cavities surrounding the guts are also fundamental in other respects to the functioning of such 'higher' animals that have transcended the architecture of sea-anemones, jellyfish and flatworms. Besides providing a medium for the rapid circulation of materials through the body, they also serve as hydrostatic balloons that can be deformed by surrounding muscles to allow finely regulated movement (as seen, for example, in the pulsed burrowing action of earthworms). Indeed, annelid and other kinds of worms, and members of several other groups that employ such cavities in their locomotion, can all be recognized in the Burgess Shale fauna.

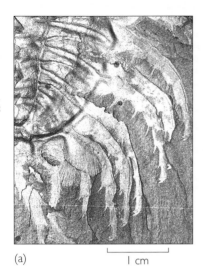

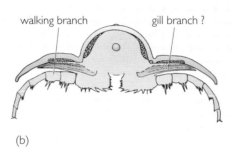

(a)   ⊢—— I cm ——⊣   (b)

Figure 3.4
(a) Detail of *Olenoides serratus*, a trilobite from the Burgess Shale.
(b) Reconstructed section across the body of *Olenoides serratus*.

Many of the skeletal elements present among the fauna were clearly also involved in locomotory systems. By far the most abundant phylum represented is the Arthropoda, which is still the most successful animal phylum today, and includes insects, spiders, crustaceans, etc. Members of this phylum are characterized by possession of an external skeleton, and by paired, jointed, limbs (Figure 3.4b).

Active burrowing into sediment and scuttling over it were thus both clearly within the competence of the Burgess Shale animals, in marked contrast to the apparently rather passive world of the Ediacaran organisms.

A proliferation of new kinds of burrows in the earliest Cambrian sediments also points to a rise in burrowing activity in general. Vertical burrows, in particular, suggest active penetration of sediment by cylindrical worms of some kind, of a grade of organization at least equivalent to those from the Burgess Shale discussed above.

If the burrow-makers were thus equipped with fluid-filled body cavities, acting as hydraulic sacs as described above, then it is likely that they possessed discrete muscular guts, as well. More widespread possession of muscular guts means, incidentally, that more organisms would probably have discharged waste material as discrete faecal pellets – a requirement of the hypothesis alluded to at the end of the last section. Faecal pellets are especially characteristic of some groups of arthropods, which package the pellets in a stiff membrane. Whether or not increased pellet production could have had the effect postulated, of enhancing organic carbon burial and retarding decomposition by heterotrophic bacteria, is another question, to which we will return in Section 3.3.2.

Along with the increase in complexity of body architecture and capabilities, the Cambrian explosion seems also to have ushered in a revolution in feeding relationships. These ecological changes could provide the key to the proliferation of skeletal hard parts.

Question 3.4
How do you suppose the worm (*Ottoia*) shown in Figure 3.3 fed?

Other indications of predation may also be seen in the fossils of the Burgess Shale. Though among the most abundant fossils in other Cambrian deposits, trilobites are usually represented in such rocks only by the mineralized articulated carapaces (external skeleton) that covered their upper surfaces (Figure 3.5a). The horny coverings of their undersides, including their limbs and other appendages (Figure 3.5b), in contrast, were not mineralized and thus hardly ever fossilized. Most trilobite fossils therefore give us rather few clues as to the feeding habits of these animals. Exceptional instances of preservation of the undersides, however, suggest a general mechanism whereby the rhythmic motion of the numerous paired limbs swept food particles into a central groove (the food channel), where they were shuffled forward by the hairy limb bases towards the animal's mouth, on the underside of the head (Figure 3.5b). The majority of trilobites probably fed in this way on organic detritus on the sea floor. One species in the Burgess Shale, however, has particularly large and robust basal segments to the paired limbs, equipped with fearsome spines (Figure 3.4: *Olenoides*). Professor Harry Whittington, the doyen of research on the Burgess Shale at Cambridge University, has interpreted these to have had a grasping and tearing function, again suggestive of predation.

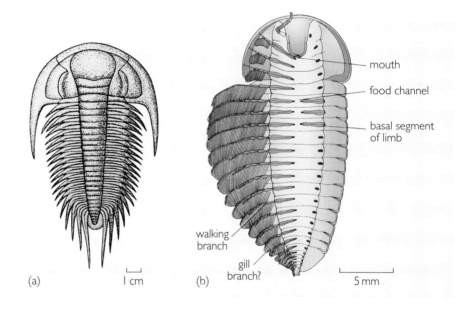

Figure 3.5
(a) Typical trilobite fossil, seen from above; (b) reconstructed trilobite underside (a different species from that shown in (a)), showing its feeding system. The full set of limbs is shown only on the left side.

mouth
food channel
basal segment of limb
walking branch
gill branch?
(a)　1 cm　(b)　5 mm

Detritus feeders are exemplified by other trilobites (as noted above), creeping molluscs and other probably related forms (Figure 3.2: *Wiwaxia*), among others. Some molluscs may also have been algal grazers.

Question 3.5
From the evidence provided so far, draw simplified trophic pyramids, like that shown in Figure 1.3, for (a) the Burgess Shale fauna, using the information given above, and for (b) the Ediacaran faunas, from that given in Section 1.5.2.

3.2.5 Hunting the hunters

So when had the revolution in feeding relationships begun? Let us start by considering the inception of predation. Particularly striking in the lowest Cambrian beds is a plethora of tiny cups, spines, scales, knobs and spicules, collectively referred to as 'small shelly fossils' (Figure 3.6a, b). Again, we owe our knowledge of these to special circumstances of preservation. Originally composed mostly of calcium carbonate ($CaCO_3$), the small shelly fossils of this age were frequently replaced, following burial, by calcium phosphate – another apparent consequence of the heightened productivity in surface waters at the time, noted earlier in Section 3.2.2. This mineral replacement means that the enclosing limestone can be dissolved away in weak acid, in the laboratory, releasing the acid-resistant small shelly fossils. Some are readily classifiable, for example, as the shells of tiny molluscs (Figure 3.6a). Others represent the scattered elements of compound external skeletons, similar to that seen in the Burgess Shale animal *Wiwaxia* (Figure 3.2). Complete assemblies of such compound skeletons were unknown from older deposits until the 1980s, when around sixty spectacularly preserved specimens were recovered from a Burgess Shale-like deposit in the Lower Cambrian of North Greenland (dating from about 530 Ma ago), covered with a 'chain-mail' coating of plates and spines (Figure 3.6c).

The plates and spines of the animal in Figure 3.6c appear to have been separately embedded in its skin, so a primary supporting function is unlikely. Despite their apparently complete coverage of the animal, they are shaped in such a way as to slide over one another, like scales, allowing free movement of the worm-like body. In this case, then, it is hard to avoid the conclusion that the main function of the coating was indeed defence, i.e. to act precisely like chain-mail. The corollary is that predatory attack was already a significant selective force, shaping the armour of such early Cambrian animals. This is not to suggest that the hard parts had themselves necessarily arisen in response to predation: selection can only act on what is already there. How they appeared in the first place remains an enigma (rather like the origin of sexual reproduction, discussed in Chapter 1).

One plausible speculation is that hard parts started as crystalline excretory products. Even the slightest degree of such mineralization in the surface tissues of an organism could have offered some selective advantage against predation, and, in the face of that selective pressure, been adapted for defence (or offence). Given the evidence for predation as a selective force in the early Cambrian, noted above, the great proliferation of skeletal hard parts at that time could readily be attributed to its influence.

For the Vendian, the evidence relating to predation is harder to evaluate. The relative scarcity of fossilized skeletal hard parts deprives us of such a telling record of either offensive or defensive features. Yet even the sparse shelly record that is available furnishes a clue. A small proportion (2.7%) of the tiny *Cloudina* shells from China have been found to contain rounded holes (Figure 3.1b), correlating in size with the diameters of the shells at the level of penetration. These holes are interpreted to have been bored by some (unknown) predator, selecting prey of a similar size to itself. Soft-bodied predators also remain a possibility, as the Cambrian predatory worm in Figure 3.3 reminds us. Indeed, jellyfish appear to have been a common component of Ediacaran faunas, and many superficially similar living forms are at least passive predators, even consuming fish and other large prey that become paralysed by the stinging cells on their tentacles. Others, however, feed instead on microscopic plankton, so we cannot be sure what the Ediacaran examples were like.

Nor is there any reason to suppose that microscopic organisms were not routinely engulfed as prey items by single-celled protists (e.g. amoebas), just as they are today. Indeed, the endosymbiotic theory for the origin of the eukaryotes (Chapter 1) relies upon this very assumption to explain the acquisition of organelles. It is

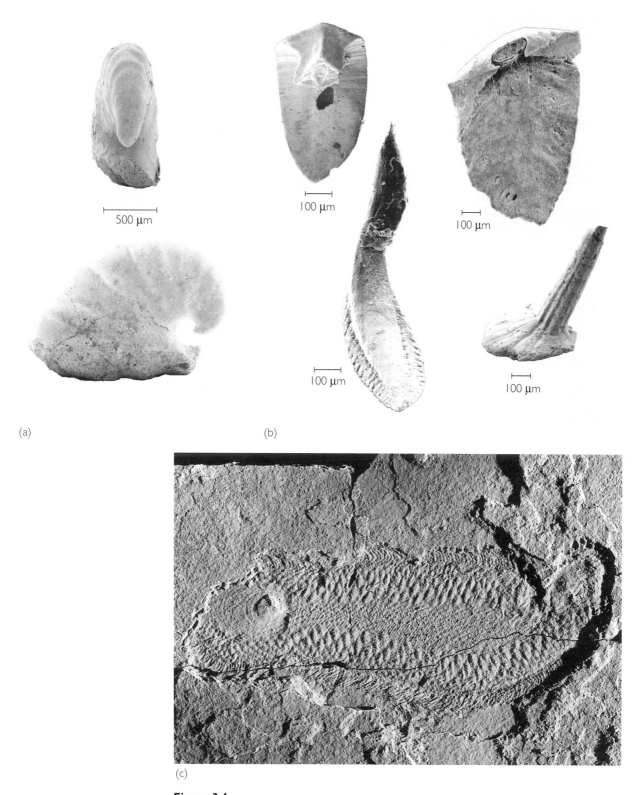

(a)

(b)

(c)

Figure 3.6
(a, b) Examples of Lower Cambrian small shelly fossils, (a) a mollusc shell, (b) isolated sclerites;
(c) complete chain-mail covering of sclerites on a worm-like animal (*Halkieria evangelista*).
The fossil in (c) is about 3.5 cm long.

probably also safe to assume that multicellular animals, such as the Ediacaran jellyfish, were also feeding at least upon microscopic prey. The possibility that some Ediacaran animals had symbiotic single-celled autotrophs in their tissues, and thus had no need to feed heterotrophically, has been mooted, but there is no evidence in favour of the idea and few support it. What is notably lacking, however, is clear evidence that any of the larger animals actively preyed upon other large animals. Note, in this respect, that both *Cloudina* and its putative predator were relatively tiny, compared with the Ediacaran animals.

In our present state of knowledge (or ignorance), it would thus appear that predator/prey relationships between macroscopic animals, at least, became important as selective influences only in the post-Ediacaran radiations: so the Phanerozoic might truly be thought of as the age of rampant consumerism! The diversification of active burrowers in the earliest Cambrian, mentioned in Section 3.2.4, is of interest in this context. The greater scope for controlled movement accompanying the acquisition of fluid-filled body cavities, which by acting as hydraulic sacs aided active burrowing, may well have opened the door for the evolution of hunting and predatory behaviour as well.

According to this scenario (which is largely informed speculation), the proliferation of skeletal hard parts that announced the dawn of the Cambrian could be regarded as an integral part of the revolutions in locomotory and feeding behaviour that accompanied the post-Ediacaran evolutionary radiations. No special environmental change need be invoked to explain the spread of skeletal hard parts at that time. Besides, the existence of some shelly forms in the late Vendian (Figure 3.1) shows that there had been no particular barrier to biological mineralization before the Cambrian explosion. If there had ever been any kind of environmental threshold for such mineralization to occur, it had already been surpassed with initially rather unremarkable effects. What seems to have changed around the Vendian–Cambrian boundary is the pressure of selection for the elaboration of skeletal hard parts.

3.2.6 Ancestry of the Cambrian animals

All this begs the question, of course, of where the newly-evolving animal groups came from. Apart from the few Ediacaran animals that might be construed as primitive relatives, the fossil record remains frustratingly mute on the issue. Certainly, the molecular evidence from the extant animal phyla, which was discussed in Section 1.4.3, suggests that their ancestors had already begun to diversify well before the Vendian. One plausible explanation for the lack of fossil evidence for the earlier ancestors of the Cambrian animals is that they were microscopic in size, perhaps dwelling in the spaces between sand grains (as many related microscopic species still do today). Unlike the larger Ediacaran organisms, whose environments they may have shared, such animals would have been unlikely to have left any impression in the sediment.

Question 3.6
Can you recall any evidence, from earlier in this chapter, which suggests that the earliest representatives of at least some animal phyla may have been very small?

Some of the ancestors of the Cambrian forms may, therefore, have been microscopic, and if, in addition, they lacked shells, they would have had little potential for fossilization. However, this seems unlikely to have been true in all cases; many of the small shelly fossils of early Cambrian age can be recognized as isolated plates from larger composite skeletons like that illustrated in Figure 3.6c. It is also likely that some of the modes of preservation prevalent at the time, such as the phosphatic replacement of calcareous shells (Section 3.2.5), may have biased the record in favour of smaller fossils; because this sort of alteration was often localized within the sediment, it tended to affect only smaller whole shells or parts of composite skeletons.

Some of the Vendian ancestors of the Cambrian animals might well already have been of large size, then, leaving us with the options either of trying to recognize them among the known Ediacaran assemblages, or of seeking their fossils elsewhere (or, alternatively, of postulating why they might not have been preserved). The issue can only be resolved by further scrutiny of the available fossils and/or the recovery of yet more informative specimens that might provide greater insights into the anatomy of Vendian animals. Lack of preserved detail is still the main stumbling block.

3.2.7 A provisional synthesis

Let us now briefly take stock of the evidence presented in this chapter, and consider again the three possible scenarios, outlined in Section 3.2.1, for the replacement of the Ediacaran faunas by those of the Cambrian.

> **Question 3.7**
> From what you have read, can you now reject any of these scenarios?

In fact the three possibilities may not be entirely mutually exclusive, depending on one's interpretation of the Ediacaran organisms. As discussed in Section 1.5.2, the Ediacaran 'faunas' probably contain quite a motley array of organisms (a few of which may yet turn out not even to have been animals). It would thus be a mistake to treat them as a distinct and coherent group, to be contrasted categorically with the Cambrian animals, as some researchers have done. All that unites them really is their common mode of preservation in Vendian sediments, although certain Cambrian survivors have been linked with them because of their specific similarities to certain Vendian forms. Perhaps it is only the relative lack of detailed evidence for their anatomy and habits that has allowed them to be lumped together. An alternative perspective suggested by Bruce Runnegar is

> ... that the Ediacara 'fauna' is merely a small sample of the Neoproterozoic [late Proterozoic] biosphere: sponges, cnidarians, bilaterians [bilaterally symmetrical animals], a possible echinoderm, and various problematical 'metaphytes' [plants] ... as well as trace fossils and worm tubes. In other words, the Cambrian explosion was well underway, but we only see some of the action through a short-lived taphonomic [preservational] window that is confined to the ultimate and penultimate stratigraphic sequences of the terminal Neoproterozoic.
> Bruce Runnegar (1995, p. 314)

From this perspective, a plausible model is that the new Cambrian arrivals radiated from just a few of the Ediacaran animals, which had acquired the key innovations (e.g. fluid-filled body cavities) that permitted the ensuing revolution in life habits. Most of the other Ediacaran forms may indeed have become extinct around the close of the Vendian, either as a consequence of some global environmental perturbation, or, perhaps, under pressure (of predation, for example) from the newly-evolving forms. Such a model would combine aspects both of the evolutionary continuity scenario and either the mass extinction scenario, or that of ecological displacement. Two things are clear, though, whatever the cause of the faunal replacement. First, the change was remarkably rapid, in geological terms. Second, it had important consequences, not just for our geological time-scale, which historically was based on the fossil record of skeletal hard parts, but also for biogeochemical feedbacks, which we will explore in the next section.

Figure 3.7 pulls together the kinds of evidence and arguments that have been discussed above, to provide a synoptic history of the inferred major faunal changes that accompanied the late Proterozoic and early Phanerozoic. The width of each balloon in the figure is intended to give an impression of changing taxonomic diversity. It is based on a synthesis published in 1993 by Simon Conway Morris, a paleontologist at Cambridge University and leading researcher on early animal radiations, but updated with the new Namibian data of Grotzinger, *et al.* (1995).

Figure 3.7
Summary diagram of early animal evolution.

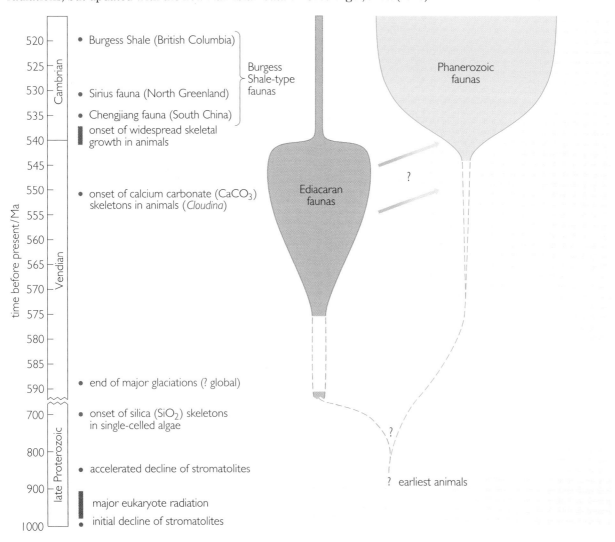

3.3 The feedback effects of the Cambrian explosion

3.3.1 Impact on the carbon cycle

The workings of the carbon cycle were explored in *The Dynamic Earth*. There it was emphasized that although the fluxes to and from geological reservoirs are tiny, compared with those of the biospheric carbon cycle, their cumulative effects in the long term have been considerable. *Atmosphere, Earth and Life* looked at the consequences of burial of organic matter for atmospheric composition. However, the proliferation of marine shelly animals that took place during the Phanerozoic saw increasing amounts of calcium carbonate ($CaCO_3$) being taken up in the growth of shells, which began to accumulate in marine sediments. As calcium carbonate is a sink for carbon, the proliferation of shelly animals impinged upon the global carbon cycle, which, as a result, took an entirely new turn.

Throughout Chapter 2, and earlier parts of this chapter, we have focused on the burial of organic matter in marine sediments: this was apparently a major sink for carbon before the Phanerozoic.

> **Question 3.8**
> What environmental condition had been particularly favourable to the burial of organic matter in Cryptozoic seas, and why might you expect that factor to have waned in importance during the Phanerozoic?

The Phanerozoic seas thus saw a decline in organic, relative to carbonate, carbon sinks. This by no means meant the end of organic carbon burial, which was to take on particular importance on land, as you will see in Chapter 4, as well as continuing at certain times and places in the sea. Nor is it to deny the importance of limestones in Cryptozoic times, which were sometimes extensively deposited. Yet before the Phanerozoic, most of the precipitation of calcium carbonate was directly from seawater, though perhaps often microbially induced. It occurred preferentially in restricted inshore and marginal environments (such as lagoons and tidal flats). Here, it is probable that supersaturated conditions, brought about by intense evaporation, allowed the free precipitation of calcium carbonate in the water, as cloudy 'whitings' of microscopic crystals (you can sometimes see a similar effect in tap water that has been boiled for too long). This would have been the main source of the carbonate mud trapped by many microbial mats to form stromatolites (Section 1.4.1) though some mat-forming microbes precipitated their own calcareous sheaths. Any deeper-water Cryptozoic limestones were generated by offshore transport of carbonate sediment produced in shallow waters.

The proliferation of shells in the Phanerozoic created an important new reservoir for calcium carbonate in open marine waters. Here they, and their debris, were allowed to accumulate to form extensive beds of shelly limestone. Hence the burial of carbon in the Phanerozoic, in both organic and carbonate form, changed radically in character, and so became subject to different influences. It is not worth attempting to quantify these changes in the relative sizes of the various carbon reservoirs from the Proterozoic to the Cambrian, because the geological record is simply too patchy to allow anything but the vaguest estimates. However, the record improves for later periods, particularly from the Mesozoic onwards, and in Chapter 7 you will have the opportunity to estimate the scale of some of these reservoirs for yourself.

3.3.2 The feedbacks of feeding

Besides generating new reservoirs, the Cambrian explosion also affected material fluxes through the diversified life habits of the animals. As noted in Section 3.2.4, there was a marked increase in burrowing activity, and the Burgess Shale (and other such deposits) shows the rich diversity of worms and other active burrowers that had evolved by mid-Cambrian times. In consequence, the surface layer (up to tens of centimetres) of marine sediments became subjected to an unprecedented level of reworking that was to continue unabated, where conditions allowed, throughout the Phanerozoic. Indeed the depths within the sediment which became churned by such activity increased as time went on, especially from Mesozoic times onwards.

The effects of burrowing were particularly important for sediments deposited in water depths below those normally stirred by waves, tides and other shallow currents. Instead of passively accumulating any organic remains that arrived, such deposits were now thoroughly churned over. Correspondingly, their contents were repeatedly exposed to degradation by heterotrophic bacteria and their consumers, and recycled to a greater or lesser extent into the overlying water (as CO_2 and H_2O). This surface layer thus became a sort of active filter between the marine and sedimentary reservoirs, modifying the flux of materials – especially carbon – from one to the other.

Question 3.9
What implication does this burrowing activity have for the hypothesis about the influence of faecal pellets on the burial of organic matter mentioned at the end of Section 3.2.3?

Since the ability to produce faecal pellets and the capacity for active burrowing were both enhanced by the same basic anatomical innovation (the appearance of internal body cavities), these two opposing factors should have increased in effect at about the same time. Their overall impact on rates of carbon burial must therefore remain an open question. However, any processes that affected the rate of organic carbon burial are also bound to have affected oxygen levels, so the combined effects of the evolution of faecal pellets and burrowing is unlikely to have been neutral.

Probably far more important, in relation to organic burial, was the degree of oxygenation of the water column itself, through which planktonic remains had to sink to reach the bottom. This presents something of a paradox in that the very process that allowed the build-up of molecular oxygen (burial of organic matter) depended upon the presence of anoxic water masses. Marine anoxia, however, did continue to develop, episodically, through the Phanerozoic, due to various combinations of factors, while the appearance of land vegetation, in due course, also added its own contribution to organic carbon burial – which is why there are Phanerozoic reserves of oil and coal, the origins of which will be discussed in Chapters 6 and 7.

3.4 Radiations and extinctions

3.4.1 Estimating the turnover of life

The Cambrian explosion is the first of the major evolutionary radiations in the history of life that can be charted in some detail from the fossil record. As was noted at the start of this chapter, this is largely because of the abundant fossil record of skeletal hard parts, especially those of marine animals. Figure 3.8 shows the total numbers of marine animal families (most containing several species, see Box 3.1) estimated to have coexisted at different times through the Phanerozoic, on the basis of fossil evidence. The compiler of this graph, Jack Sepkoski of the University of Chicago, USA, limited it to the marine fossil record because this is relatively more complete than that of land-dwelling organisms.

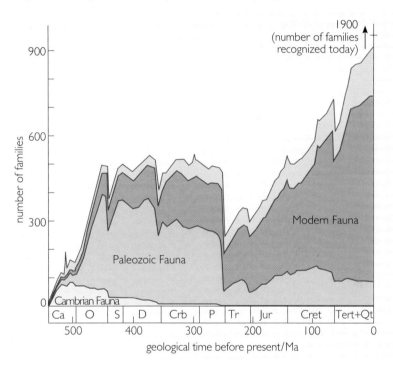

Figure 3.8
Changes in the numbers of families of marine animals through the Phanerozoic, based on fossil evidence, according to Jack Sepkoski. The pink area at the top represents those families known only from instances of exceptional preservation. The main, blue area below is based on the fossil record of skeletal hard parts, and is comprised of three 'evolutionary faunas', which are explained in Section 3.4.3.

▪ What might explain the contrast in relative completeness of the marine and land-based fossil records?

▪ Most sediment ultimately ends up in the sea, so sedimentary deposition there is relatively more widespread, and less interrupted, than on land.

A later compilation, by Mike Benton of the University of Bristol, UK, covering all organisms, is shown in Figure 3.10a, for comparison. Figures 3.10b and c, respectively, show the continental (land-dwelling) and marine components of that compilation. For all three graphs in Figure 3.10, minimum and maximum estimates of family numbers are shown, which allow for uncertainties over the durations of many families in the record, and the representation of some fossils in both marine and continental deposits. Notice that Figure 3.10c differs in detail from Figure 3.8, reflecting a mass of little differences in the family classifications used, and in the estimates of their stratigraphical ranges; these differences provide a useful reminder that such information is not a fixed set of facts, but the result of

Box 3.1 The hierarchy of classification

Organisms, both living and fossil, are classified into a nested series of increasingly inclusive groups, known as **taxa** (sing. taxon), all the way up to the level of the kingdoms discussed in Box 1.1.

This system, known as the **taxonomic hierarchy**, is ideally intended to reflect evolutionary relationships, though where these are unclear an essentially pragmatic scheme has to be adopted, which can be subject to correction in the light of new findings. Taxonomic categories at different levels in the hierarchy are given standard names (e.g. Family), each taxon being assigned its own latinized name (e.g. Canidae, the dog family). The taxonomic categories employed for animals are shown in Figure 3.9, along with the classification in this system of the domestic cat and dog.

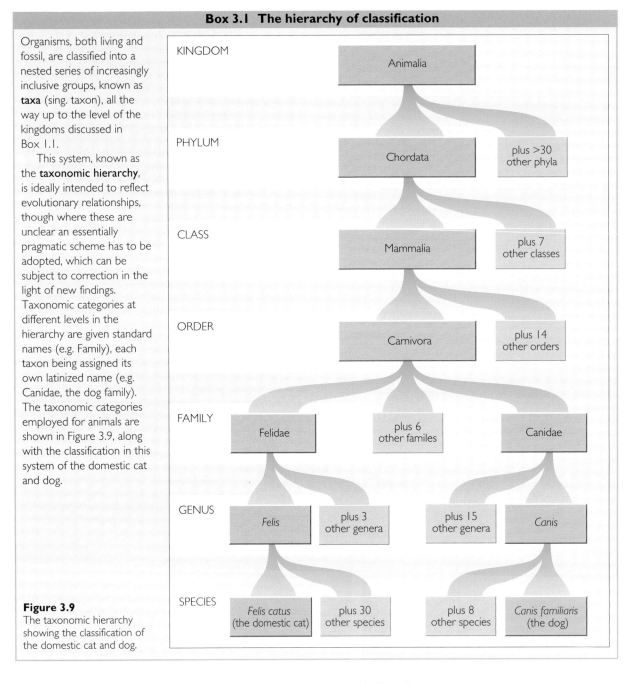

Figure 3.9
The taxonomic hierarchy showing the classification of the domestic cat and dog.

interpretation both in the classification of fossils and in the correlation of sedimentary sequences. The record is always subject to revision with new finds and new analyses of existing data. The similarity of the two marine compilations nevertheless suggests an underlying pattern that is not obscured by such differences of interpretation.

You may wonder why numbers of species have not been directly plotted, instead of families (in both compilations). After all, families are artefacts of classification, while species are natural biological units. The reason is that this is the simplest way to cope with the incompleteness of our knowledge: fossils record but a fraction

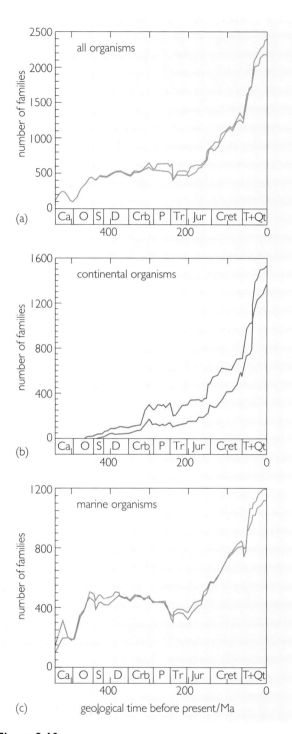

Figure 3.10
Changes in the numbers of families through the Phanerozoic, according to Mike Benton: (a) all organisms; (b) continental (land-dwelling) organisms; and (c) marine organisms. Maximum and minimum estimates are shown. Reprinted with permission from Benton. J.: 'Diversification and extinction in the history of life', *Science*, **268**, p. 53. Copyright 1995, American Association for the Advancement of Science.

of past species, only some of which have been described and classified. Being more inclusive, families have a relatively more complete record. One specimen alone from one species in a family is sufficient to record the family's presence. By analogy, a telephone directory would provide you with a fairly complete dossier of all the households in a town, but a great deal more effort would be required to account for all the individuals. Families are thus a useful proxy for numbers of species. Of course there is a risk of biased representation because families are of very different sizes. The working assumption, however, is that such inequalities average out, or at least still allow broad patterns at the species level to be revealed, when large enough samples are considered.

The main, blue, area under the curve in Figure 3.8 represents the record derived from skeletal hard parts only (ignore, for the time being, the subdivision of this area into three components). For the sake of completeness, the pink zone along the top shows the additional record of families known only from exceptional fossils of animals that lacked durable mineralized skeletons.

Question 3.10
What do you suppose the sharp peak in the pink zone, in the middle part of the Cambrian, represents?

Such instances provide a sobering reminder of what is missing elsewhere, but, precisely because they are exceptional, they have to be excluded from consideration of how diversity has changed in relative terms. The skeletal hard part fossil record represents only a fraction of life, but it at least provides a consistent basis for monitoring the relative ups and downs in family numbers, by allowing us to compare like with like from different times. Hence the following analysis relates only to that record, i.e. the blue area in Figure 3.8.

The Cambrian explosion is marked on Figure 3.8 by the steep rise in family numbers through the early Cambrian. Throughout this episode the evolution of new families (i.e. first appearances of species with their diagnostic characters) clearly outstripped extinctions (i.e. final disappearances of constituent species). The pattern has been likened to what happens to numbers of individuals in a population when the birth rate exceeds the death rate.

▓   Can you think of some examples of the lattercircumstance?

▓   The most obvious example is that of the human population explosion, itself a
     consequence of a decrease in the individual death rate (i.e. increased
     survivorship of individuals), due to social and medical advances, without a
     globally matching decrease in the birth rate. If you are a gardener, you may
     also have thought of the unwelcome boom in aphids each summer: this is
     unleashed by the onset of rapid asexual reproduction, easily outstripping the
     death rate of individuals, as they gorge on the new growth of their plant
     hosts.

As for the individuals in the analogies above, the 'birth' of new species and
eventually new families, through the divergent evolution of populations, must have
been virtually unrestrained during the early Cambrian. That so many evolving
populations should have thrived in myriads of habitats, with but a small toll of
extinctions, is certainly consistent with the idea of a post-Ediacaran 'ecological
vacuum'. This is not to say that competition was virtually absent, as is sometimes
mistakenly suggested in this context. As noted at the start of this book (Section
1.3.4), fitness differences between individuals *within* populations, hence competition
between them, is why evolution by natural selection occurs. What is suggested,
instead, is that during the early Cambrian there was an abundance of ecological
opportunities for evolving populations to exploit without risk of exclusion by
incumbent species: there was 'all to play for'. Rather than being less intense, then,
competition was probably less constraining, resulting in rapid evolutionary
divergence.

3.4.2 Mass extinctions

Diversity did not increase indefinitely. The data in Figure 3.8 suggest a temporary
lull in late Cambrian times, though diversification resumed spectacularly in the
Ordovician. The reasons for this lull are still unclear, though some regional
extinctions, especially of trilobites, have been documented in North America. Much
more striking, however, is the sharp drop in family diversity (number of families)
accompanying the close of the Ordovician. This marks the first of five large-scale
Phanerozoic mass extinctions to which the fossil record testifies (not to mention a
sixth in which we are implicated today).

▓   What was the approximate percentage decrease in total numbers of families
     recorded from fossil hard parts at the close of the Ordovician?

▓   Total family numbers fell from around 470 to about 340 (remember to take the
     figures from the line at the top of the blue area, and ignore the pink area
     relating only to exceptionally preserved assemblages). Hence the percentage
     decrease in family diversity was $[(470 - 340)/470] \times 100\% = 28\%$.

Two things need to be said about this result. First, it represents only the net decline
in family numbers. Given the appearance of some new families over the interval,
the percentage *extinction* of pre-existing families would have been somewhat
higher. Second, the values of percentage extinction at lower taxonomic levels – i.e.
of genera and species – would have been yet higher still. Consider a family that lost
all but one species: it is still counted as having survived, although at the species
level there may have been a considerable loss, which must be added to the toll of
species extinctions in families with no survivors. By sampling patterns among
well-documented groups at lower taxonomic levels, it is possible to estimate such

losses. By using this method, it has been estimated that some 85% of species may have become extinct at the end of the Ordovician.

Question 3.11
Identify the times of the other four great mass extinctions that followed, from the graph in Figure 3.8.

Greatly increased rates of extinction over more, or less, limited periods of time are implicated in all of these mass extinctions, and also in a number of other examples of smaller scale. However, a temporary suppression of the rate of speciation, and hence origination of new families, is also evident in some cases, at least. By far the most severe mass extinction was that at the end of the Permian, when more than half (57%) of the families of marine animals became extinct. The estimated toll at the species level was a staggering 96%. We will delve further into the possible causes of this disaster in Chapter 6, where we can view it in the context of global change at that time.

A postulated link between the late Cretaceous mass extinction (including that of the dinosaurs) at the close of the Cretaceous and a probable major asteroid impact, signalled in the sediments by a sharp peak in content of the rare element iridium, has generated much debate and popular interest. In 1984, Jack Sepkoski and a leading researcher on mass extinctions, David Raup, also from the University of Chicago, undertook a detailed analysis of the distribution in time of all detectable mass extinctions. They concluded, for the post-Paleozoic examples at least, that they appeared to show an approximately 26 Ma periodicity, and offered the daringly imaginative interpretation that this could reflect a regular cycle of extraterrestrial bombardment.

A vigorous debate ensued, with opponents criticizing several aspects both of the database and of the analysis, particularly the recognition of a periodic signal, and the proponents claiming confirmation of their result from more detailed analyses. That there have been asteroid and cometary impacts on the Earth, some of considerable size and effect, is not in doubt (as shown in *Origins of Earth and Life*); several forms of evidence, ranging from the preserved impact craters themselves to deposits of beads formed from volatilized rock, shocked quartz, and distinctive minerals formed only under very high pressures, as well as tell-tale geochemical signatures (such as sharp increases in iridium content) in associated sediments, provide ample evidence of their past occurrence. What is open to question, however, is the extent to which impacts have been responsible for mass extinctions, in general, and, in particular, whether mass extinctions really show a periodicity that might reflect astronomical 'forcing'. At present the jury remains out on the latter question, with critics arguing that the observed pattern does not deviate significantly from what can be produced by random coincidence in a model involving numerous unrelated kinds of environmental perturbation (including occasional impacts).

Evidence linking some of the other extinctions of the 'big five' to impacts is subject to debate, though such evidence has also been reported at a couple of younger (Tertiary) levels, associated with more modest mass extinctions. Moreover, the apparent durations of the various mass extinctions, and their relative effects on different groups of organisms, were by no means consistent from one example to another. Hence it is likely that a variety of causes may need to be invoked to explain them, and the geological record offers several compelling Earth-bound candidates (as detailed in Box 3.2).

Box 3.2 The 'big five' Phanerozoic mass extinctions

In detail, many of the mass extinctions show a more complex pattern than what is revealed in Figure 3.8.

The main episodes of mass extinction shown by the Phanerozoic fossil record, together with their main casualties, and likely causes, were as follows:

1 Late Ordovician

Two main peaks of extinction, towards the close of the period, were separated by hundreds of thousands of years. Both plankton (e.g. graptolites) and bottom-dwelling life (e.g. trilobites and reef-building organisms) were affected. Associated events were the growth and decay of a vast ice sheet on the southern supercontinent of Gondwanaland (the assembly of which was discussed in Section 2.2.1), as it moved over the South Pole, with drastic effects on sea-level, climate and ocean chemistry.

2 Late Devonian

The pattern of extinction is as yet unresolved, but possibly comprised a series of extinctions extending over at least 3 Ma, with the most severe effects some 5 Ma before the end of the period. Shallow marine ecosystems were most affected, with tropical reef-dwellers being particularly hard-hit. Temperature fall again seems to be implicated, associated with widespread anoxia in shallow seas. Although there is no direct evidence for glaciation then, sea-levels fluctuated and fell overall. A positive shift in $\delta^{13}C$ values in the C_{carb} record would be consistent with burial of organic carbon and consequential drawdown of CO_2. On the other hand, some evidence for impacts is provided by glassy fragments from Belgium, as well as impact swarms in Chad and North America.

3 Late Permian

Increased extinction rates occurred over some 3–8 Ma at the close of the period, although recent findings suggest that these fell largely in two distinct episodes. Marine organisms were particularly devastated (Figure 3.8). A complex, probably synergistic, array of causal factors, including biological feedbacks, has been postulated (as will be discussed in Chapter 6).

4 Late Triassic

There were at least two, maybe three, extinction peaks during the last 18 Ma or so of the period. In the sea, both free-swimming animals, especially ammonoids and marine reptiles, and bottom-dwelling forms, again including many reef-building organisms, declined. On land, many reptilian groups, including mammal-like forms, were lost, as well as large amphibians and many insect families, although there was no marked global extinction of land plants. The marine extinctions, at least, coincided with marked changes in sediment type, strongly suggestive of major climatic change, and, in particular, much more extensive rainfall has been suggested. Nevertheless, a huge impact crater in Quebec, some 65 km across, dates from around the Triassic–Jurassic boundary. It is known as the Manicouagan structure. Only very low levels of iridium have been found in the melt rocks associated with the impact, so tracing its 'geochemical signature' in sedimentary rocks elsewhere, to test for correlated extinction, may prove difficult.

5 Late Cretaceous

Two patterns appear to have been superimposed: several groups of marine animals, both free-swimming (e.g. belemnites and some marine reptiles) and bottom-dwelling (e.g. certain kinds of bivalves), seem to have dwindled over the last 9 Ma or so of the period. At the end, however, there was an abrupt collapse, especially amongst the plankton. This happened within perhaps 100 000 years and maybe much less, and is marked by fluctuations in the carbon isotope record of marine limestones. Ammonites combined both patterns of extinction, with a slow decline terminated by the rapid extinction of a sizeable number of the remaining species at the end of the period. On land, the most famous demise is that of the dinosaurs, though the pattern of their decline is still debated. They were joined, too, by the pterosaurs (flying reptiles). Flowering plants suffered major losses at the end of the period, especially at mid-latitudes in North America (where carbon isotopic data from organic residues indicate mass mortality) and at high latitudes in Asia. In the Southern Hemisphere, however, changes were gradual or non-existent. A number of other groups were also little affected, including, for example, crocodiles, snakes and placental mammals. While the final collapse is widely interpreted as impact-related (with evidence mounting for the impact crater itself being buried in the subsurface of the Yucatan Peninsula, Mexico), it evidently occurred in ecosystems already perturbed by other causes, as the more gradual background decline shows. The latter can be related to substantial changes in continental configuration, climate and oceanic circulation at the time.

One thing that is nevertheless clear is that all five major collapses of diversity were brought about by environmental crises of one sort or another. There is no evidence that the rapid diversification of any groups of organisms directly provoked any of these past mass extinctions (unlike that which we are in the process of implementing today). Each mass extinction seems to have come as a shock to the system, cutting sharply across any pre-existing pattern of change in diversity. And each one re-set the agenda for subsequent evolution. The next section explores how.

3.4.3 Evolutionary radiations: opportunity knocks

Look again at Figure 3.8, and in particular, the periods immediately following each of the 'big five' mass extinctions shown there.

> **Question 3.12**
> What happened to diversity in the immediate aftermath of each of these extinctions? Suggest a reason for the pattern shown, on the basis of what you have read so far in this chapter.

So much for the immediate consequences of the mass extinctions, but was there any longer-term trend?

▒ What happened to overall diversity levels in the longer term?

▒ Overall, diversity has risen, though not in a continuous fashion. After the Ordovician, an approximate 'plateau' of diversity was established, until the late Permian mass extinctions. Thereafter diversity continued to rise, overall, until the present day.

Throughout the Paleozoic, after the Ordovician, the recovery of diversity following the mass extinctions appears to have tapered off each time to similar levels, of between 400 and 500 families. There has been much debate about whether or not the Paleozoic 'plateau' of diversity levels might reflect some kind of evolutionary equilibrium, with the post-Paleozoic trend signifying growth perhaps to a new, higher, equilibrium level. The implicit assumption of this idea is that increased diversity leads to increased rates of extinction and decreased rates of origination, as a consequence of increased competition between species.

> **Question 3.13**
> How could such a relationship between diversity and rates of extinction and/or origination yield an equilibrium in diversity?

It remains open to question, however, whether the rate of extinction ever did catch up with that of origination (other than during mass extinctions) to yield the postulated equilibrium levels. An alternative interpretation is that the Paleozoic 'plateau' of diversity may just be the net effect of several major and minor mass extinctions (the latter below the level of resolution of Figure 3.8), superimposed upon a background pattern of sustained diversification. That is the explanation preferred, for example, by Mike Benton, to explain the patterns in Figure 3.10. Possible controls on the rates of origination and extinction of taxa, and hence on global diversity, are still vigorously debated.

Whatever the long-term influence of the extinctions on the numbers of families over time, they certainly altered the composition of life in the sea. This effect is revealed by the three bands indicated within the blue area of the graph in Figure 3.8. These represent sets of major groups (mainly classes) of marine animals, each set showing a characteristic pattern of family turnover. Jack Sepkoski referred to these sets as **evolutionary faunas**, and their separate diversity histories, as well as some representatives of the major groups assigned to them, are illustrated in Figure 3.11. The first set, referred to as the Cambrian Fauna, diversified much more rapidly during the initial radiations of that period, than did the others. However, it soon began a long decline in numbers of families as the next, Paleozoic Fauna began burgeoning. The latter continued diversifying until the late Ordovician, after which it too began a slow decline, while the third, Modern Fauna continued to expand alongside it, eventually dominating the scene in post-Paleozoic times.

▨   Do you recognize any of the types of animals depicted in Figure 3.11?

▨   From the Cambrian Fauna, you should have recognized the trilobite and the primitive mollusc (monoplacophoran), both of which were illustrated earlier (Figures 3.5 and 3.6a, respectively). You might have recognized, too, the conical shell of the hyolithid, as the unfortunate victim of the predatory worm, *Ottoia*, shown in Figure 3.3. Several of the types of fish and 'shellfish' from the Modern Fauna might also be familiar to you from general knowledge, as perhaps also some of the fossil forms in the Paleozoic Fauna. Don't worry, however, about those that are unfamiliar to you: the intention of the illustrations is to convey an idea of the different kinds of animals comprising each evolutionary fauna.

Note, incidentally, that the names of these evolutionary faunas refer only to their times of dominance: all three have existed throughout the Phanerozoic (although very few members of the Cambrian evolutionary fauna survived beyond the Paleozoic). Moreover, the three faunas do not comprise evolutionarily discrete sets of animal groups. In some cases, different classes from a single phylum have been allocated among different faunas. The molluscs, for example, are split among the three, with bivalves (clams) and gastropods (snails) in the Modern Fauna, cephalopods (including, especially, the ammonoids) in the Paleozoic Fauna, and the primitive monoplacophorans in the Cambrian Fauna.

Sepkoski himself explained these broad patterns of change in terms of a complex version of the equilibrium diversity model outlined above. He postulated that each successive fauna (for whatever unknown reasons) had its own characteristic diversity-dependent rates of origination and of extinction of families, and thus its own intrinsic equilibrium diversity level. The details of how this model could explain the histories of the three faunas shown here, and the arguments for and against it, need not concern us. Nevertheless, the response of these evolutionary faunas to mass extinctions is worth dwelling on.

Question 3.14
Did the mass extinctions affect all three evolutionary faunas equally?

Thus each mass extinction, particularly that at the close of the Permian, left a biased set of animal groups among the survivors. The reasons for such differential survival are not well understood. They need not necessarily just reflect differences in the fitness of individuals in their perturbed environments, according to the

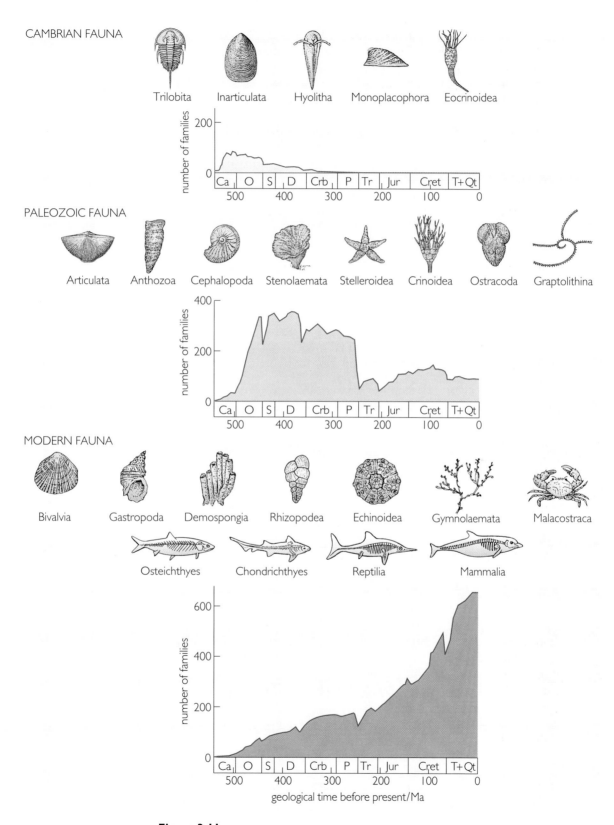

CAMBRIAN FAUNA

Trilobita   Inarticulata   Hyolitha   Monoplacophora   Eocrinoidea

PALEOZOIC FAUNA

Articulata   Anthozoa   Cephalopoda   Stenolaemata   Stelleroidea   Crinoidea   Ostracoda   Graptolithina

MODERN FAUNA

Bivalvia   Gastropoda   Demospongia   Rhizopodea   Echinoidea   Gymnolaemata   Malacostraca

Osteichthyes   Chondrichthyes   Reptilia   Mammalia

geological time before present/Ma

Figure 3.11
Examples of representatives, and diversity histories, of the three evolutionary faunas of skeletalized marine animals indicated in Figure 3.8.

normal principles of natural selection (Section 1.3.3). It is also possible that certain properties of whole groups, such as their geographical distribution, may have had a hand in their survival; there is some evidence that genera containing species distributed in many different regions stand a better chance of surviving a mass extinction than those with species all clumped within one region. This form of hierarchical filtering can readily be explained by the environmental perturbations affecting different regions unequally (see Box 3.2).

The ensuing radiations of course built on the characteristics of the survivors. The complexion of animal groups coexisting at any given time has therefore been fashioned quite as much by the devastations of extinction as by the achievements of adaptive evolution. But what were the consequences of these major faunal changes, in ecological terms?

3.4.4 Ecological relationships and their consequences

Some associated ecological changes among marine organisms are indeed apparent. Perhaps the most remarkable of these changes again involved predation, and it followed on from the late Permian mass extinction. As you saw earlier (Section 3.2.4), there was, even in the Cambrian, no shortage of predators, some even ingesting entire small shelly prey. Yet, throughout the Paleozoic, there was an abundance of exposed shelly animals that lived rooted to the sea-floor (some examples are shown amongst the Paleozoic Fauna in Figure 3.11). Evidence for predation on them, in the form of diagnostic damage preserved in the fossil shells (similar to that shown in Figure 3.1), however, is relatively infrequent, although a distinct increase in the late Paleozoic has been recorded. Throughout Mesozoic times, however, the fossil record shows evidence for a further marked increase in the intensity of predation on shelly prey. The evidence is not restricted to the direct record of damage inflicted by predators. Both the appearance of many new kinds of predators with specialized adaptations for tackling shelly prey, and the emergence in shelly prey species of a host of new defensive adaptations, can be detected (Figure 3.12).

Among the newly-evolving predators in Mesozoic seas, for example, were crabs and lobsters, as well as several groups of fish and marine reptiles, variously equipped to crush, smash or pierce shells. New kinds of starfish evolved the ability to pull open bivalved prey and insert their stomachs, to digest the occupants. In the Cretaceous these were joined by gastropods capable of drilling through shells to reach their prey, and somewhat later, in the Tertiary, by various shell-breaking birds and mammals.

The major groups of shelly prey animals showed a wide variety of responses. One notable evolutionary trend was a boom in burrowing, especially by bivalves and echinoids (sea-urchins), to increasing depths within the sediment, in refuge from the predators. The churning of the surface sediment, mentioned earlier in relation to burrowing (Section 3.3.2), was thus both intensified and deepened. This effect, coupled with an increase in the extent of disturbance at the surface by roving grazers and detritus feeders (sometimes colourfully referred to as 'biological bulldozing'), as well as the onslaught of predators, in turn made ever more hazardous the larval settlement of shelly animals that grew permanently anchored to the sediment surface. The latter, which had been so widespread in the Paleozoic, now waned in relative diversity.

These linked ecological changes have been collectively referred to by the American evolutionary biologist, Geerat Vermeij, as the **Mesozoic Marine Revolution**.

Figure 3.12
Reconstruction of the predators and prey of a shallow Cretaceous sea floor in southern England. In the section of burrowed sediment in the foreground, a burrowing clam (bivalve) is being attacked, at the centre, by a shell-drilling snail (gastropod), while a burrowing sea-urchin (echinoid) penetrates the sediment on the left. In the middle distance, on the left, a starfish tackles a mussel (attached to the kelp), while a regular echinoid grazes on small sessile colonial animals also growing there. To the right, mussels and oysters (both bivalves) are being attacked by another drilling gastropod, a crab and a lobster (the last two, arthropods). Various shell-crushing fish hover in the background, and the jaw of one (a shark), with its pavement of flat teeth, lies on the surface in the centre foreground. Old bivalve shells are scattered around, that in front of the crab showing a gastropod drill-hole.

They transformed the character of marine life to that which can still be recognized today. It could well be that this revolution also helped to provoke the unprecedented rise in marine animal diversity after the Paleozoic, noted earlier.

Perhaps the most profound feedback to biogeochemical cycles came from associated changes in the plankton. For, starting in the late Triassic, groups of microscopic plankton with calcareous skeletons began to appear, including both single-celled algae (e.g. coccolithophores) and protists (e.g. planktonic foraminifera). It is tempting to interpret this simultaneous adoption of calcareous skeletons (or toughened organic walls in some other planktonic groups) as a defensive adaptation against increased grazing pressure.

Question 3.15
What effect might such calcareous plankton have had on the distribution of carbonate sediments?

Such offshore deposition again led to amplification of the carbonate part of the carbon cycle, particularly from Cretaceous times, when these planktonic groups underwent major radiation. We will return to these effects later, in Chapter 7.

Going back again in time, however, some quite different evolutionary changes during the Paleozoic had had even more considerable environmental consequences – perhaps the most profound, indeed, since the oxygenation of the atmosphere way back in the Proterozoic. These were the changes that allowed plants and animals to invade the land, and to which we turn in the next chapter.

3.5   Summary of Chapter 3

1   The geological record for the late Vendian suggests a number of environmental upheavals, involving, for example, a global fall in sea-level, and marked fluctuations in the rate of burial of organic material (hence climatic instability) followed by lowered rates in the Cambrian. Nevertheless, Ediacaran faunas persisted to the end of the Vendian, reaching maximum diversity in the final 6 Ma, and a few forms even survived into the Cambrian. It is unclear whether the majority of Ediacaran animals suffered a mass extinction at the close of the Vendian, or whether they were ecologically displaced by newly-evolving animals.

2   Exceptionally preserved fossil assemblages of soft-bodied animals from the Cambrian reveal their anatomical advances over the earlier Ediacaran animals. Many of the Cambrian forms show greater differentiation of body parts, including the concentration of food-trapping organs around a 'head' end, the appearance of limbs and of discrete tubular, two-ended, guts. Most significant was the associated appearance of fluid-filled spaces within the body, which allowed the gut to be decoupled from external body movements, and which could serve as hydraulic sacs giving new scope for finely regulated movement.

3   Accompanying these anatomical changes seems to have been a revolution in feeding relationships. In particular, the rapid proliferation of skeletal hard parts may reflect the rise of predation, and hence of multi-tiered trophic pyramids.

4   The proliferation of shells from early Cambrian times impinged upon biogeochemical cycles, leading, for example, to increased deposition of limestones in offshore open marine environments. Phanerozoic seas thus saw a shift in emphasis from the burial of carbon in organic material (with the decreasing frequency and extent of oceanic anoxia) to that of carbon in carbonate rocks. Increased burrowing in offshore sediments also helped to reduce the extent of organic carbon burial there.

5   The fossil record for families of marine animals with hard parts provides a synoptic guide to Phanerozoic mass extinctions and radiations. The early Paleozoic radiations are consistent with the idea of there having been a relative ecological 'vacuum' then. Subsequent diversification was interrupted by a succession of mass extinctions, of which five were notably severe, coming in the late Ordovician, the late Devonian, the late Permian (the most devastating), the late Triassic and the late Cretaceous.

6   Although periodic extraterrestrial impacts have been proposed for these, and other smaller mass extinctions, differences in the relative durations and the effects of the extinctions, together with other geological data, point instead to a mixture of Earth-bound and extraterrestrial causes.

7   Each mass extinction was followed by a relatively rapid rebound in family numbers. In the longer term, however, there was an overall increase in diversity levels, though whether this involved successively higher equilibrium levels, or merely the effects of a dynamic interplay between ever rising diversity and numerous extinction events, remains unresolved.

8   Within the pattern of diversification of marine animals, three 'evolutionary faunas' may be distinguished: the Cambrian Fauna dominated the initial radiations, but then tailed away thereafter; the Paleozoic Fauna continued diversifying through the Ordovician, but then commenced a long slow decline;

while the Modern Fauna more slowly, but relentlessly, expanded, eventually rising to dominance after the Paleozoic. Each successive fauna seems to have been less drastically affected by mass extinctions than its predecessor. The extinctions thus left somewhat biased line-ups of survivors.

9   There was an intensification of predation, especially upon shelly prey, and most notably expressed among post-Paleozoic faunas. One major defensive adaptation appearing among the prey animals was deep burrowing into the sediment, which led to yet further churning of surface layers. These, and other linked changes, are collectively referred to as the Mesozoic Marine Revolution. The associated rise of various planktonic groups with calcareous skeletons in the Mesozoic augmented carbonate sedimentation in deeper water, so further expanding the oceanic carbonate sink for carbon.

Chapter 4
Greening of the land

4.1 Introduction

As you have seen earlier in this Course, land vegetation interacts with the atmosphere in various important ways – mainly through photosynthesis and respiration, which both involve gas exchange and the cycling of water vapour. In effect, vegetation acts as a major processor of the atmosphere.

To put the importance of the role played by land vegetation in perspective, consider the relative contributions to net primary production that are made by the land and the sea. Net primary production, as you may recall from *The Dynamic Earth*, is the rate at which biomass is produced by photosynthesis *minus* the rate at which it is used up during respiration by the primary producers themselves. The total world net primary production of biomass is $1.70 \times 10^{14}\,kg\,yr^{-1}$ (expressed here in terms of the actual amount of biomass, rather than in terms of the quantity of carbon incorporated into living tissue); a huge proportion of this, $1.15 \times 10^{14}\,kg\,yr^{-1}$ or approximately 67%, is produced by land plants, even though land represents less than one-third of the global surface area. Two factors are involved here: first, land plants are more efficient at carrying out photosynthesis, and secondly, the actual amount of living plant biomass at any one time is much larger than the amount of living marine biomass because, on the whole, plants, and in particular trees, live longer than photosynthetic marine organisms and produce complex multi-layered communities of great bulk. Thus, at any given time, the total land-plant biomass is about five hundred times greater than the total marine biomass and a thousand times greater than that of all the plankton in the oceans. Significantly, over 90% of the world's non-bacterial biomass is in forests.

> **Question 4.1**
> Given that all photosynthetic organisms need nutrients (most of which are ultimately derived from the weathering of rocks) and light in order to grow, suggest why so much more photosynthetically active biomass is sustained on land than in the sea.

In the global carbon cycle (see *The Dynamic Earth*), land vegetation is by far the most important biospheric reservoir of carbon. It also provides food and habitats for animals, and it significantly affects the heat balance of the planet through its influence on albedo. In this chapter, we shall examine how land vegetation evolved and what effect that evolution had on Earth and life (Sections 4.3 – 4.7). First, however, it is useful to consider briefly land vegetation today (Section 4.2), as this will provide a reference point in interpreting the past.

4.2 The power house of the biosphere

Land vegetation is predominantly composed of green plants, and green plants constitute the 'power house' of the biosphere. Underlying almost all trophic pyramids (Section 1.3.2) is the process of photosynthesis, which, as you will recall, utilizes the energy in sunlight to bring about the combination of carbon dioxide and water to form organic molecules (sugars or, more generally, carbohydrates), with oxygen being liberated as a by-product. In simple terms, the rate at which carbon is fixed (i.e. taken out of the atmosphere as carbon dioxide and incorporated into organic molecules) by this process depends on three things: the concentration of carbon dioxide in the atmosphere, the availability of water and nutrients, and the flux of light energy at usable wavelengths. Since the evolution of land plants between 400 and 500 Ma ago, the most variable of these factors have been the concentration of atmospheric CO_2 and the availability of water.

To illustrate just how sensitive the present-day interrelationship is between the level of CO_2 and the amount of plant growth on a global scale, consider Figures 4.1 and 4.2. (Note that although the periods covered by Figures 4.1 and 4.2 are not identical, the pattern in each is repeated year by year.) Figure 4.1 shows the annual fluctuations in atmospheric CO_2 concentration at different latitudes. You can see that the largest fluctuations are in the Northern Hemisphere where most land area and, therefore, most vegetation exist. In the Southern Hemisphere, which is mostly ocean- or ice-covered, the fluctuations are small and, as would be expected, six months out of phase with those in the North. Now, compare this with Figure 4.2 which shows the amount of 'greenness' (measured in terms of the relative absorption of a standard wavelength in the visible spectrum) over time and latitude, greenness being taken as a proxy for 'leafiness' and hence for the amount of photosynthesis. Again, the largest fluctuations are in the Northern Hemisphere, though greenness in the Tropics is relatively constant. Another, though not unexpected, feature is the low level of greenness in the desert latitudes (25–35°), but this is a little harder to see on the figure. Most scientists regard spring–summer growth (and the consequent take-up of CO_2) and autumn–winter dormancy (coupled with winter fossil fuel combustion) as being primarily responsible for the CO_2 fluctuations, with the maximum CO_2 levels occurring during the winter months.

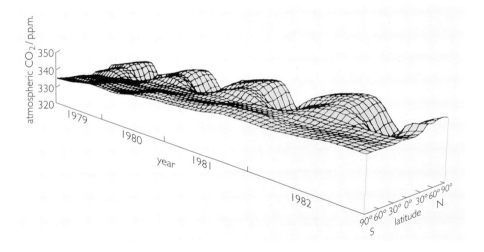

Figure 4.1
Annual fluctuations in atmospheric CO_2 concentration plotted against latitude.

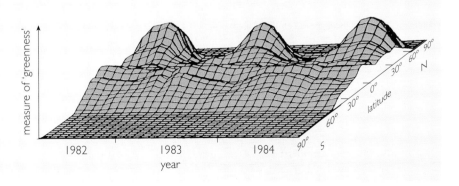

Figure 4.2
Annual fluctuations in the amount of 'greenness' (due to vegetation) plotted against latitude.

Clearly, plants are accustomed to modest variations in atmospheric CO_2 concentrations over short time-scales and can adapt to them. Larger variations can produce more noticeable effects; for instance, it is well-known that crop plants grown in a CO_2-enriched atmosphere are more productive than those grown under similar conditions but with a normally prevailing atmospheric CO_2 level. However, experiments show that the effects of such CO_2 enrichment are often short-lived, and it is questionable what the long-term consequences might be.

It is also difficult to extrapolate back in time in order to deduce what the effects of different CO_2 levels might have been in the distant past. Indeed, it is hard for us to imagine what the Earth was like when climate, atmospheric composition (particularly CO_2 content), vegetation and animal life were very different from those of our present world. Understanding what the land environment was like before the advent of land plants, at a time when life was essentially aquatic, is especially difficult. But that is exactly what we have to do if we wish to appreciate the impact that the invasion of the land by plants had on the Earth.

4.3 Before the invasion by plants

4.3.1 An alien world

Imagine you have travelled to an alien planet. The atmosphere is similar to that of the Earth, but there is less oxygen and breathing is difficult. Nevertheless, you do not need a spacesuit. You are standing on a rocky promontory overlooking a large desert plain – a desert that is far more harsh than any on Earth today. Large areas consist of nothing but bare rocks, some sedimentary, others volcanic. A mixture of colours confronts you – white, yellow, grey, but with reds and blacks predominating – visually harsh because there are no greens, no plants. A dry valley to your left has a network of intertwining gullies leading to a lake shimmering beneath a cloudless sky. Ringing the lake is a wide band of salt crust. To your right and stretching away to the horizon are sand dunes. There are no birds in the sky – no animals at all. The only sounds that you hear are those of wavelets lapping on the lake's edge and the hiss of sand grains borne on the wind. You feel the Sun burning your skin, and you are very thirsty.

This alien planet is, in fact, Earth, but we have travelled back in time to about 500 Ma before the present – the third imaginary journey you took at the beginning of Chapter 1 (Question 1.1). This hostile world is about to be transformed into one that is far more familiar.

4.3.2 The land without plants

The landscape seems to have been very bleak 500 Ma ago, but let us consider why it was so. Without vegetation to bind them, the surface sediments produced by rock weathering, as well as by volcanism (lava and ash), chemical precipitation and crystallization (salts etc.), would have been highly mobile. Winds would have created sand and dust storms, forming large dune fields in places; moreover, the storms themselves would have eroded rock surfaces, yielding more sediment. In fact, there would have been little soil; what there was, would have been largely limited to wet habitats where cyanobacterial populations could thrive. Elsewhere there would have been just a covering of loose sediment, giving the land a desert-like appearance overall.

Another factor contributing to this desolation can be inferred from modern deserts. One noticeable feature of these deserts is the pronounced variation in temperature

(a)

(b)

Figure 4.3
(a) A meandering river.
(b) A braided river.

over a daily cycle. During the day, solar radiation is usually unhindered by clouds, and the ground surface heats up, often to temperatures of more than 50 °C. At night, those same cloudless skies allow the heat to leak rapidly out to space in the form of long-wavelength radiation, and temperatures at ground level may fall below freezing. This daily temperature variation causes thermal expansion and contraction stresses which shatter rocks, thereby providing debris for mobilization by wind and water. These processes would also have operated on the Earth 500 Ma ago, adding further to the bleakness of the landscape.

Evaporation from the oceans would have cycled some water into the atmosphere, only to fall later as rain. But, for reasons we shall come to shortly, the average rainfall would have been less than now, though with some places nevertheless consistently wetter than others. Wind and water would have transported the loose sediments to the oceans with no hindrance from rooted vegetation. River flow would have been highly variable, and flash floods common – a consequence of rainwater running directly off rock or clay surfaces, as well as groundwater flowing easily through sands to emerge in springs and rivers, rather than being 'mopped up' by soil. This highly variable flow regime would not have produced gently flowing meandering rivers like that shown in Figure 4.3a, but rivers where the channels divided and rejoined many times, as in Figure 4.3b, and where flow rates varied greatly. Such rivers are called 'braided', because they look like intertwining braids of hair. They form because, as flow wanes, the sediment being transported chokes the channels and the residual flow is forced to split into a multitude of streams.

The sedimentary record confirms that rain-fed lakes and rivers existed before the arrival of land plants. However, without vegetation, the amount of water cycled through the atmosphere, particularly over the continents, would have been much less than at present, and hence there would have been fewer clouds and the rainfall would have been lower. A modern example of this effect is provided by the Amazon Basin: about half of the rainfall there is due to humidity produced by evaporation from the plants themselves; consequently, cutting down the forest has resulted in measurable drying of the climate in that area.

Without land plants, there would have been no food for animals and therefore no terrestrial animal life. Moreover, some two-thirds of the photosynthetic capacity of today's Earth would have been missing, and so the oxygen content of the atmosphere would have been lower than now. It is unclear how much less effective the ozone shield might have been as a result (see *Atmosphere, Earth and Life*), but the desert land surface might have suffered increased levels of ultraviolet radiation, as well as searing daytime heat.

Not only were the land conditions much more harsh than today, but the terrestrial environment presented the invaders with certain problems not encountered in the oceans. We shall now consider these problems and the consequent adaptations that allowed the move on to the land.

4.4 From ocean to land: making the change

4.4.1 Problem 1: the fight against gravity

An aquatic plant has a density only marginally greater than that of the water that surrounds it. Small amounts of gas (the gaseous products of photosynthesis and respiration) in the tissues of the plant provide positive buoyancy. The plant therefore does not have to expend energy in building rigid support structures against the pull of gravity (although it does need tensile strength to resist the action of the waves). That is why seaweeds are limp when left high and dry on a falling tide. Land plants, by contrast, need to invest energy and resources in building structural support.

4.4.2 Problem 2: bodily fluids

In an aquatic environment, a plant is bathed in a nutrient solution. Hence, seaweeds, for example, have no obvious fluid conduction system for the transport of water and dissolved substances around the plant: each cell is close enough to the water surrounding the plant body for the nutrients and water to diffuse in and out. On land, mineral nutrients and water are only available on a regular basis from the soil, so a plant of any appreciable size has to have a way of distributing the nutrients and water from the soil to all of its cells. It therefore needs a fluid distribution, or **vascular**, system.

To exchange water, nutrients and metabolic waste products, the outer skin of a seaweed has to be permeable to these substances. A seaweed on land would dry out quickly (think how dry and crispy seaweed gets when it is washed up on a beach and exposed to the Sun). Land plants have evolved a barrier between their cells and the external atmosphere – a protective layer, or **cuticle**, often covered in waxes. Fossil evidence suggests that one of the key adaptations for the successful exploitation of the terrestrial environment was the evolution of the cuticle.

The cuticle covering also means that land plants have a very limited capacity to absorb water directly from the atmosphere. Instead, those parts of the plant in direct contact with the land surface are specially adapted to absorb water and soluble minerals from the substrate, as well as to provide anchorage. A seaweed, on the other hand, needs only a part specialized for anchorage.

> **Question 4.2**
> Why do you think some aquatic plants, such as water lilies, have stems with vascular systems, a cuticle on their photosynthetic surfaces, and roots that are adapted for absorption as well as for anchorage?

4.4.3 Problem 3: reproduction

If an organism is not to be overcrowded by its progeny, the offspring must move away from the parental site. In water, a ready transport system for both gametes (sex cells) and offspring is provided by currents. Although land plants can exploit the wind in a similar way, the gametes and progeny are susceptible to drying out and hence to death. Some land plants, like ferns, exploit intermittent wet periods to spread their gametes and achieve fertilization, and so they might be regarded as only partially adapted to land. Others, such as flowering plants, the **angiosperms**, can effect fertilization in the driest of conditions and thus show the most advanced adaptations to dry land.

While desiccation is a major problem affecting all aspects of a land plant's life, it is most threatening during reproduction, with the gametes and young embryos being especially vulnerable. This problem would have been most acute during the initial colonization of the land when conditions were extremely harsh. Of course, the problems of desiccation cannot be separated from the associated issues of structural support and nutrient supply which confronted the first land plants and which we considered above, nor from the related issue of gas exchange which we address in Section 4.5.

4.4.4 Adaptations and consequences

Land plants evolved in such a way as to invest energy and metabolic products in building structural support and developing specialized structures for anchoring into loose soils and for absorbing and distributing water and nutrients from those soils. Plants did this while maximizing gas exchange and, at the same time, limiting water loss. Such a complex set of interdependent features is usually built up in stages, each step being an adaptive success under the prevailing circumstances, but one that opened up possibilities for further adaptations. It is likely, then, that these adaptations, which had the overall effect of maintaining an aquatic environment within the plant, did not evolve all at once. Yet, within the context of the geological time-scale, they arose rapidly.

Not only did plants retain the aquatic environment within themselves as they moved onto land, but they also modified the external environment in a way that proved to be more benign to them and to all the animal life dependent on them. In particular, the arrival of plants on the land profoundly altered the hydrological cycle there and the associated biogeochemical processes.

4.5 How is water cycled through the atmosphere by vegetation?

In this section we shall focus on the role played by plants in the hydrological cycle.

As you know, photosynthesis powers the biosphere: energy in sunlight brings about the combination of carbon dioxide and water to form sugars and oxygen. In photosynthesis, land plants use the light in the blue and red parts of the spectrum, and they reflect the other, mainly green, visible parts of the spectrum, which is why they appear that colour. The basic reaction is:

$$6CO_2 + 6H_2O + energy \longrightarrow C_6H_{12}O_6 + 6O_2$$

Question 4.3
What is the reverse of this biochemical process among living organisms?

Both photosynthesis and respiration involve gas exchange. In an aquatic environment the gases are usually dissolved in water, but in a terrestrial environment the exchange is directly with the atmosphere. Land plants therefore must have a mechanism for allowing gaseous carbon dioxide and oxygen to diffuse in and out of the plant body. However, in this case it is not practical for the exchange to take place over the entire plant surface as in an aquatic plant. Why is there this difference? The key factor here is that water molecules are small enough to pass through cell walls that are permeable to carbon dioxide and oxygen. So, if gas exchange in a land plant occurred over the whole plant surface, there would be considerable accompanying water loss which would result in the desiccation of the plant and its subsequent death.

A simple solution to the gas-exchange problem in land plants would be to have holes in an otherwise impermeable membrane covering the plant body. In fact this is what we see in some early land plants where there are pores in the cuticle which covered the plant (Figure 4.4).

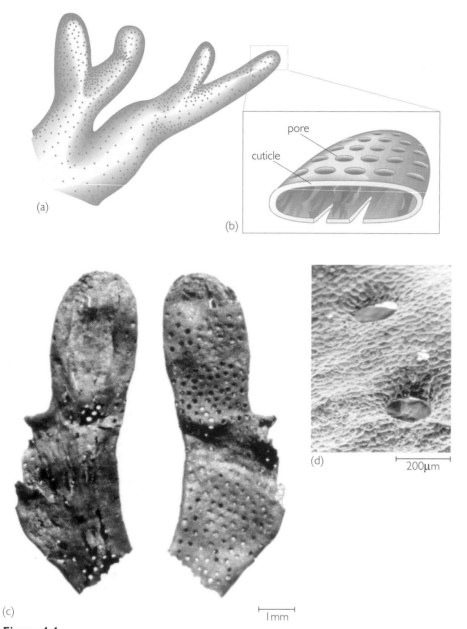

Figure 4.4
(a) *Spongeophyton*, an early land plant reconstructed from fossil specimens, viewed from above.
(b) Details of the growing tip, shown in section. (c) Photograph showing an actual specimen of
Spongeophyton, with, on the left, the underside and, on the right, the top side of a growing tip.
(d) Scanning electron micrograph of the pores seen in (c).

These simple pores would have provided no regulation of water diffusion, but by
the early Devonian, land plants had evolved microscopic controllable openings,
termed **stomata** (sing. stoma). A single stoma in its basic form consists of two
guard cells, each of which is shaped like a slightly bent sausage with its concave
side facing its partner (Figure 4.5). When the guard cells have a lot of water in
them, their curvature increases so a hole opens up between them. When they

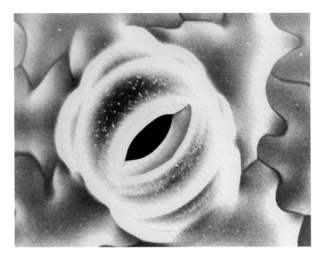

Figure 4.5
False-colour scanning electron micrograph of an open stoma on the surface of a leaf of the tobacco plant, *Nicotiana tabacum*. The two guard cells are readily visible. (Magnification: × 850)

contain only a little water, they are straighter and the hole closes. In this way, when the plant can afford to lose some water, the guard cell aperture opens and gas exchange can then take place for photosynthesis. By contrast, when the plant is parched, water conservation is a priority, so the guard cell aperture closes, inhibiting gas exchange and, therefore, photosynthesis. In most plants the aperture also closes in the dark, when photosynthesis would not be possible.

Question 4.4
What typical attribute of living organisms does the regulatory device provided by stomata exemplify?

This simple yet effective regulatory system appears to have been very successful in minimizing the water loss problem in land plants, and stomata are still possessed by almost all such plants.

Even so, the gas exchange associated with photosynthesis inevitably leads to some water loss through the stomata; indeed, considerable quantities of water enter the atmosphere by this means. This water loss to the atmosphere is known as **transpiration**. Perhaps surprisingly, it is of some benefit to the plant. This is because root-to-leaf water flow is maintained by the continual loss of water from the parts of the plant above the ground, and so transpiration enhances the transport of dissolved salts within the plant body. Evaporation from the surface of the leaves also helps to cool plants in hot, dry weather when overheating could damage or slow down important biochemical reactions.

Another way, besides transpiration, in which plants contribute to high levels of water vapour in the atmosphere is by intercepting rainwater on their leaf surfaces, from where it evaporates back into the air without ever touching the ground. Now, if all the leaves on a plant were laid out side by side, they would typically occupy an area many times greater than the area of ground covered by the plant itself. This ratio of leaf area to basal ground area is known as the **leaf-area index**. For grassland it is typically between 1 and 2, for temperate woodland, between 4 and 6, and for a tropical rainforest, more than 7.

In a forest, much of the water falling though the leaf canopy is intercepted by leaf after leaf as it falls earthwards. It therefore wets a larger area of leaves than that of the ground it would otherwise hit. Because the rainwater is spread over a large area, it evaporates rapidly back into the atmosphere and less of it enters the soil/groundwater or river systems. So, not only is water cycled through the atmosphere more rapidly and in larger volumes when plants are present than when they are not, but groundwater flow is moderated. Chemical weathering of rocks, erosion and sediment transport to the sea are all consequently affected.

Groundwater is also extracted by the root systems of plants and passes through the leaves to the atmosphere by transpiration, again reducing the amount of water available for immediate water flow in the ground and in the rivers. However, the higher atmospheric moisture content produced by transpiration leads to higher rainfall. Thus, water is circulated more often and more rapidly in the evaporation–rainfall cycle, while water leaking from this cycle into the groundwater and river systems does so in a more gradual fashion. As a consequence, river discharges are more even from catchments covered in vegetation than from those devoid of plants. This, in turn, contributes to a change in river geometry from a braided to meandering form.

The spreading of plants from the aquatic environment to what had previously been desert therefore had many consequences and was one of the most significant steps in the evolution of life on Earth. It is worth examining this step in some detail because, to a large extent, it determined the subsequent pattern of life on land, including the evolution of human beings.

4.6 The land turns green

4.6.1 Beginnings

Although it is generally accepted that the beginning of the colonization of the land by green plants took place in the Silurian or even late Ordovician, there is evidence that bacterial mats and, possibly, fungi may have been prolific enough on damp land surfaces to have formed primitive soils as early as the late Proterozoic. This is not hard to imagine because there is a present-day parallel: the sandy desert floors of the southwest of the United States are partially stabilized against erosion by crusts incorporating cyanobacteria. The early soil-forming microbial communities would have locally altered the colour of the landscape and incorporated some organic matter into the land substrates, but otherwise they would have had only a minor effect on global systems compared with the overall changes that took place during the Silurian and Devonian.

The first convincing evidence that plants had begun to adapt to a dry-land environment comes in the form of microscopic fragments of tube-like structures, cuticle and spores (Box 4.1).

Box 4.1 What is a spore?

Reproduction in plants follows a somewhat modified version of the route shown earlier in Figure 1.5. Crucially, meiosis in diploid plants gives rise to **spores**, instead of producing gametes directly (Figure 4.6).

A spore consists of a living cell surrounded by a tough water- and chemical-resistant outer coat. The cell is haploid, and develops into a haploid individual which, in turn, produces gametes by mitosis. Fusion of two gametes then yields a new diploid individual that goes on to produce a new generation of spores by meiosis.

The spore coat protects the haploid generation during dispersal and, because the coat is so tough, the spore itself has a high probability of being preserved in the fossil record. Many sedimentary rocks contain thousands of fossilized spores in each cubic centimetre, and the different spore contents of the rocks allow stratigraphic correlation.

Pollen grains are specialized male spores in which the development of the haploid generation is highly abbreviated: the sperm cells are released directly and then fuse with egg cells to produce seeds. In this way, seed plants have effectively curtailed the haploid generation in favour of the development of diploid individuals.

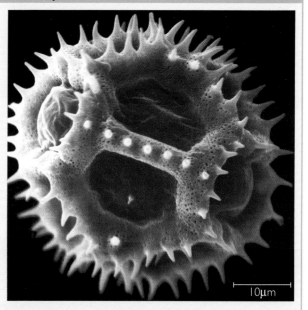

10μm

Figure 4.6
Scanning electron micrograph of a specialized male spore, or pollen grain, of chicory (*Cichorum intybus*). The complex ornamentation (i.e. the spikes on the surface) aids in dispersal by insects and can be used for spore identification.

4.6.2 Interpreting the structure of early land plants

The move from an aquatic to a terrestrial environment was accompanied by fundamental changes in plant architecture. Initially, land plants were low-growing sheet-like structures that existed almost entirely within a relatively still and humid layer of air next to the ground – the so-called **boundary layer**.

It is useful to consider the boundary layer in a little more detail. Suppose a wind is blowing (or air currents are in motion), and free air, i.e. air well above the ground surface and unimpeded by trees, buildings etc., is moving relative to the ground at a certain speed. Close to the ground, the air will be moving much more slowly because of the slowing effect of friction (viscous drag) due to the ground surface. The wind speed will therefore decrease progressively in the vicinity of the surface, so immediately above the ground there will be a relatively still layer or 'skin' of air – the boundary layer (Figure 4.7).

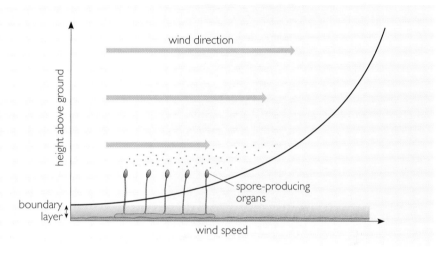

Figure 4.7
Variation in wind speed with height above a surface over which air is flowing. Superimposed on this plot is a diagram showing the height and characteristics of a typical early low-growing land plant, such as *Sporogenites*, which has only its spore-producing organs projecting above the boundary layer. The superimposition of the two diagrams could be interpreted as suggesting that the plant 'stalks' on the left of the diagram are exposed to a lower wind speed than those on the right; however, this would be wrong, as the intention is just to show the relationship between boundary-layer thickness and plant height.

The thickness of the boundary layer will depend on the speed at which the free air is moving and on the roughness of the surface. Typically, the thickness of the boundary layer over a sand or gravel surface will be of the order of a few centimetres. If the ground surface is moist, then this layer of relatively still air can become saturated with water vapour. This means that the boundary layer usually has a higher relative humidity than the free air above it (unless, of course, all the air is water-saturated).

A plant growing along the ground within the moist boundary layer would tend not to dry out because of this high humidity. Moreover, it would not need to expend energy and resources building 'plumbing' systems, enhancing its mechanical strength, or making cuticles. Yet, the lack of flowing air or water around the plant body in the boundary layer would be disadvantageous for dispersing progeny. By contrast, any plant capable of growing up into the moving air above the boundary layer would not have this problem (Figure 4.7), and so taller early land plants would have had an advantage in the race to colonize the land surface.

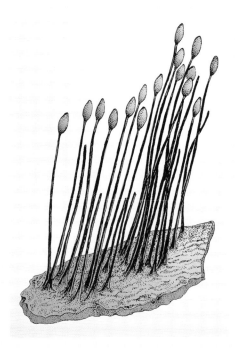

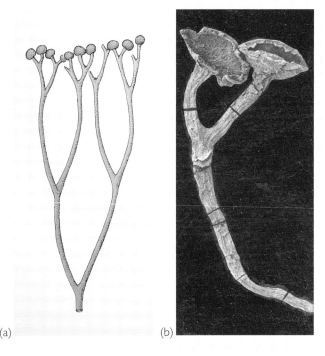

(a)   (b)

Figure 4.8

Sporogenites exuberans, an early Devonian plant (about 2 cm tall), reconstructed from fossil specimens. The pouches that contained the spores are clearly visible at the tops of the stalks.

Figure 4.9

(a) Reproductive units (about 7 cm long) of *Cooksonia caledonica,* a late Silurian plant, reconstructed from fossil specimens such as that shown in (b). (The specimen in part (b) is 4 mm tall.) No specimen of the rest of the plant has been found.

Accordingly, in early land plants, the ground-hugging body of the plant remained in the boundary layer, while stalks with special pouches containing the spores grew upwards (Figure 4.8). These spore-producing organs exploited the moving air currents above the boundary layer to carry the spores away on the wind to colonize new territory, and if the spores were small enough, they could travel great distances quickly.

Reproductive units are preferentially preserved in the rock record, and the presence of a cuticle enhances their potential for preservation. Many fossils of land plants comprise only isolated reproductive units (Figure 4.9b), while the rest of the plant remains unknown.

◾   How might the thickness of the boundary layer have been affected by the increasing vertical height of early plants?

◾   The presence of taller plants would have, in effect, increased the roughness of the ground surface and hence the thickness of the boundary layer.

This positive feedback loop would have promoted vertical growth because the thicker boundary layer would have meant that the favourable moist growing conditions were extended upwards, encouraging greater projection of spore-producing organs. But with vertical growth would have come an increase in the cost of construction, because greater structural strength would have been required and, in addition, a plumbing system would have been needed to move fluids up (water) and down (sugars produced by photosynthesis) the plant body. This increase in construction cost would, in turn, have necessitated more food production – in other words, more photosynthesis. More photosynthesis would have required more gas exchange with the atmosphere, more plant mass and therefore a larger surface area to intercept light. Pretty soon, there was branching, shading, and yet more vertical growth in plants.

Many plants became adapted to minimize their energy expenditure while growing tall. All common early land plants and representatives of all the groups of taller plants in the Silurian and Devonian shared a number of similarities (Figures 4.10 and 4.11). All had a prostrate stem, known as a **rhizome**, with small root-hair-like appendages (**rhizoids**) that anchored the rhizome to the substrate, and all had vertical stems sprouting up from the rhizome at intervals (see, for example, Figure 4.10a and b). The horizontal rhizome branched out over the substrate, so a single plant might occupy several square metres with numerous vertical stems for reproduction. Thus, in an early Devonian landscape there would have been patches of vegetation composed entirely of extensive thickets no taller than a few tens of centimetres.

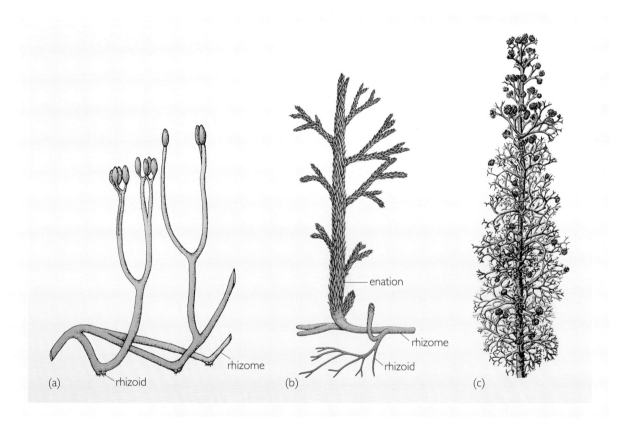

Figure 4.10
Reconstructions of some common early land plants, showing various structural features.
(a) *Aglaeophyton major,* early Devonian (about 50 cm tall). (b) *Asteroxylon mackiei,* late early Devonian (about 30 cm tall). (Enation was a special kind of outgrowth from the stem, which functioned as a small leaf.) (c) *Pertica quadrifaria,* early Devonian (about one metre tall overall).

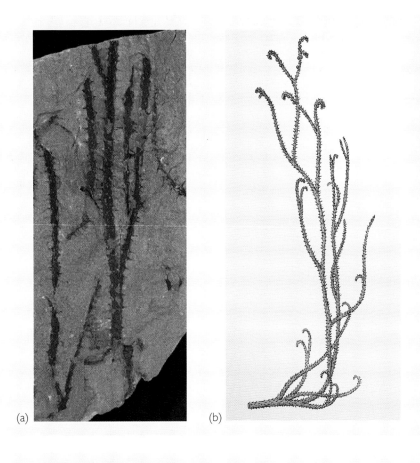

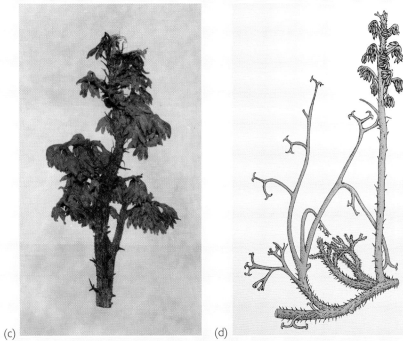

Figure 4.11
Fossil and reconstruction of (a) and (b) *Sawdonia ornata*, early Devonian (about 50–90 cm tall); and of (c) and (d) *Psilophyton dawsonii*, early Devonian (about 50 cm tall).

The vertical stems of these early plants were often branched and, at or near the apex of the vertical stem, there were spore-producing sporangia – the evolutionary 'business end'. The stems bore no leaves, but many had hooked or spine-like outgrowths on them (see, for example, Figure 4.10c and Figure 4.11b and d). These outgrowths were quite small and poorly supplied with fluid transport tissue. Their role in increasing photosynthetic area was probably minimal. Their curled or hooked form suggests that one of their functions was to enmesh neighbouring stems, thereby providing mutual support. Upright growth was also aided by the fact that the crowding of the stems increased the thickness of the boundary layer. A modern-day parallel for this crowding effect is seen in a field of wheat: a single stalk has very little mechanical strength and, if exposed on its own to the wind, it would quickly be blown over. Many stalks growing close together, however, can survive upright throughout an entire growing season.

To summarize, surface roughness and hence the thickness of the boundary layer increased as land plants evolved, their growth driven upwards by reproductive advantage and mutual shading (Figure 4.12). Beneath the 'canopy' of vegetation, a humid environment was maintained, in which the need for thick cuticles was reduced.

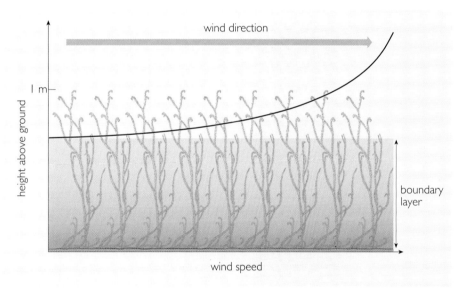

Figure 4.12
Sawdonia heath, with an increased boundary-layer profile. As in Figure 4.7, two diagrams are superimposed here: one shows the height and characteristics of the plant and the thickness of the boundary layer, and the other shows the wind speed in relation to the height above the ground. The same caveat applies as in Figure 4.7.

The simple, leafless architecture of the earliest land plants tells us little about the climate of the time. The naked branching stems with limited vascularization (plumbing) are like those of modern desert plants. This simplicity would have allowed the plants to grow in a wide range of environments; indeed, each genus known from Devonian sediments has been found over several continents. Remember, though, we can only define fossil species from morphology; we cannot tell if they were genetically related.

■ Why is it important to state that the species are defined by their morphology?

■ Because species alive today are usually defined (at least, in sexually reproducing organisms) in terms of their capacity to interbreed freely with other members of the same species but not with members of different species. In fossils, particularly those of primitive plants, we do not have good data on breeding limits in the original plants, so we have to define species purely by their morphology. Similar-looking plants might, in fact, have been genetically distinct.

4.6.3 The evolution of the leaf

By the late Devonian, multi-layered forest ecosystems were fully developed, populated by plants with broad flattened leaves with high leaf-area indices (Section 4.5). Evaporation and transpiration, and hence the cycling of atmospheric water over land, must at last have approached present-day levels. Although the earlier Devonian heaths like those of *Sawdonia* would have had a major impact on erosion and sedimentation rates through the binding action of the rhizoids and rhizomes, their effect on the hydrological cycle would by no means have been as great as that caused by the advent of leafed plants. Without the evolution of the leaf, our present world could not exist.

A leaf is generally thought of as being a flattened organ that produces food for the plant by means of photosynthesis. Commonly, leaves are green; even when they are not, it is usually because the green colour of the light-trapping pigment, chlorophyll, is masked by other pigments. Leaves come in a variety of shapes and sizes, and their architecture varies with the species and the environment. We shall return to the environmental constraints on leaf architecture later (Chapter 7), as fossilized leaves provide a powerful tool for determining past climates, but first we need to consider why the leaf evolved at all and why it is found in the vast majority of terrestrial plants.

Take a leaf from any common plant – preferably one that has leaves which you can see through. What do you notice? One of the most obvious features is that most leaves are very thin – often not more than a fraction of a millimetre thick. This thinness ensures that all the cells in the leaf are close to the atmosphere with which the plant has to exchange gases. Of course, some leaves are quite thick, but this thickness tends to be a specialization that increases the plant's ability to conserve water, as the plant's surface area to volume ratio will be reduced. Some leaves may also have thicker than usual coverings of cuticle or waxes, which also enhance water conservation. Overall, the size, shape and thickness of a leaf is always a compromise between conflicting demands – for example, between maximizing light capture and minimizing water loss.

Another notable feature of leaves is the network of veins in them (Figure 4.13). Usually the network consists of a **midvein**, or **primary vein**, which gives off a series of less thick **secondary veins**. These, in turn, have thinner **tertiary veins** running between them, and sometimes even finer orders of veins criss-cross the leaf until very small areas of the leaf (areoles) are enclosed by veins.

The veins have two functions: (i) they are the plumbing system of the leaf, supplying water and nutrients and distributing the carbohydrate products of photosynthesis to the rest of the plant; and (ii) they provide structural support for a web of photosynthetic tissue that sometimes exceeds a square metre in area. The branching pattern of veins is reminiscent of the branching pattern of a tree. This similarity is not altogether accidental, for it appears that almost all the leaves we see around us today, whether they be of an oak tree or a fern, were derived from modified branches. The only exceptions are the leaf-like enations of some primitive plants such as club mosses or *Asteroxylon* (Figure 4.10b), which are outgrowths of the stem.

So, how did leafed plants evolve? Take a look at Figure 4.14. It attempts to bring together what is known about the evolution of the plant characteristics that appear to have been significant in increasing the gas-exchange, carbon-fixation and water-cycling processes in early land plants. The important message of the diagram is

Figure 4.13
The vein system of a leaf.

that there appears to be a broad correspondence between the drawdown of atmospheric CO_2 and the innovations in land-plant architecture. As time progressed, plants increased in size, simple naked branched stems developed a single main stem with side branches, and these side branches became flattened and webbed with tissue to form large photosynthetic surfaces (leaves). This increase in leaf area demanded greater fluid movement in the plant body so vascular systems became more complex and better developed. In turn, this provided more fluids and nutrients to the leaves which became even larger allowing more carbon to be fixed and providing more carbohydrates with which to build larger plants. This positive feedback loop was eventually moderated when atmospheric CO_2 concentrations became limiting in the context of temperature, water availability and other factors and when the structural costs of building larger plants became prohibitive, if not in absolute terms, then in competitive ones.

You will see that, in the Devonian, atmospheric CO_2 was estimated to have been at least nine times higher than at present (as discussed in *Atmosphere, Earth and Life*). If true, this fairly high concentration of CO_2, coupled with the relatively small biomass of plants, would have meant that CO_2 was unlikely to have been a limiting factor for photosynthesis at that time. With primitive vascular systems, and less water than now being cycled through the atmosphere, water rather than CO_2 may have limited photosynthesis in a large number of plants. Interestingly, in many early Devonian plants, stomata, which in the absence of leaves would have been scattered over the stems, were rare compared with their prevalence in today's plants and even in plants of the Carboniferous.

Question 4.5
Why do you think early Devonian plants could have functioned with only a few stomata?

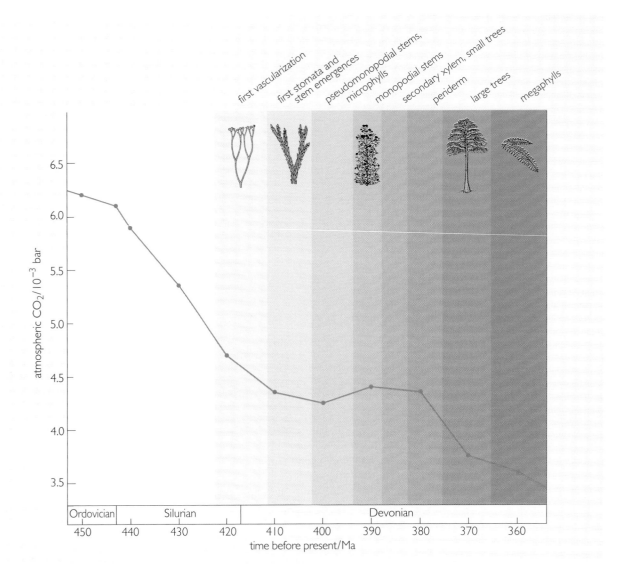

Figure 4.14
Major innovations in land-plant architecture with respect to the atmospheric CO_2 level. Pseudomonopodial stems refer to a single vertical stem made up of unequal branches, while monopodial stems denote a genuine main stem. Microphylls are small outgrowths of the stem, while megaphylls are true leaves. Secondary xylem is the tissue that makes up wood, while periderm is rather like bark.

As colonization of the land proceeded, more and more carbon was being fixed and transferred from the atmosphere to the soil, with increasing amounts ending up buried in sediments. Land plants therefore added a major new factor to the biological drawdown of CO_2, which, prior to their evolution, had taken place only in the marine realm.

With atmospheric CO_2 decreasing, it, rather than water, eventually became limiting for photosynthesis, and plants adapted by increasing their stomatal numbers. But, the presence of stomata, and the fact that cuticles and waxes are not totally impervious to water, inevitably led to greater loss of water from plants to the atmosphere. To compensate for this, more water had to be supplied to the leaf. This, in turn, led to natural selection for more efficient root systems and better vascularization within plants. Thus, with increasing branching systems and leafy photosynthetic areas, we see a corresponding increase in vascular complexity in late Devonian plants. Alongside this, the diameters of the water transport cells (tracheids) also increased in many taxa (Figure 4.15). The early ferns provide some good examples of these developments (Figure 4.16).

Figure 4.15
Changes in tracheid diameter in land plants over time.

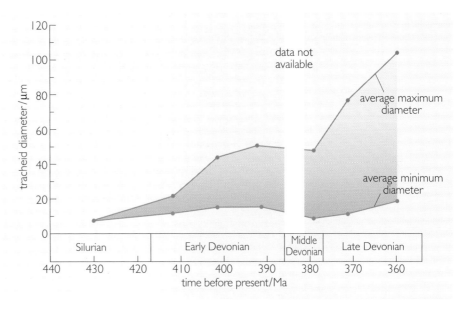

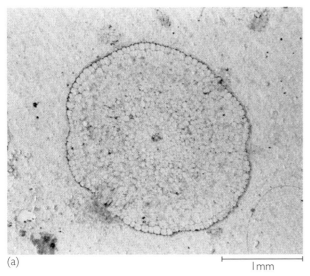

(a)

|—————|
1 mm

Figure 4.16
Early plants and their vascular systems. (a) Cross-section through a stem of the Early Devonian *Rhynia* plant, with the dark patch in the centre being the single vascular strand. (b) Cross-section through the stem of a Carboniferous fern, *Psaronius*, showing a far more complex vascular system with many arching worm-like strands in the central region (the circular structures around the edge are cross-sections through roots).

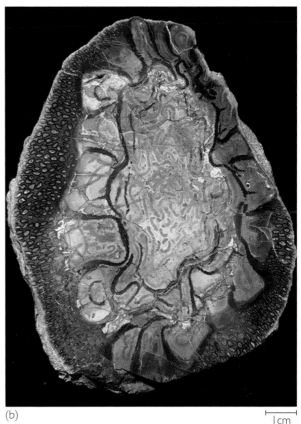

(b)

|—————|
1 cm

The evolution of leaves in the late part of the Devonian greatly increased the surface area for gas exchange and also for the evaporation of water vapour into the atmosphere, thereby causing the atmosphere to cycle water at a greater rate. The evolution of leaves also contributed to the cycling of carbon because many long-lived plants discard their leaves and replace them several times during their lifetimes, which enhances carbon transfer from the atmosphere to the soil and the sediments. The innovation of leaf loss and replacement appears also to have evolved in the Devonian. As the mantle of rock waste became increasingly enriched with such organic matter, so rates of chemical weathering were stepped up (as will be explained in Chapter 6), and rich soil profiles spread across the land. Thus, the carbon, water and other biogeochemical cycles are inextricably linked together by the photosynthesis of land plants.

4.6.4 The first forests

The late Devonian saw a major diversification in land plants. In fact, most of the main groups of plants, with the exception of the flowering plants (the angiosperms) and a group called the cycads (to be discussed in Chapter 7), appear to have had their origins in the late Devonian. One of these early groups, the **progymnosperms**, was especially important as it represented a further increase in carbon sequestering.

The progymnosperms were the dominant canopy formers in the earliest forests. The story of their identification is of interest because it provides an example of why we cannot afford to be too restricted in our concepts of ancient organisms – it can be misleading to interpret ancient plants only in terms of plants that are living today. The progymnosperms produced, for the first time, wood that had all the characteristics of the wood of modern conifers (which, together with other primitive seed-bearing plants, are referred to as **gymnosperms**). However, the progymnosperms also had foliage that looked like that of a modern fern and which bore reproductive structures that produced not seeds, as in modern conifers, but spores, like a fern (Box 4.1). As long as the wood and foliage were found unattached to one another, it was thought that they came from quite different plants, and they were classified in different groups. The wood, given the generic name *Callixylon* (Figure 4.17a), was classified as belonging to the gymnosperms, while the foliage, which was referred to the genus *Archaeopteris*, was thought to belong to the ferns (Figure 4.17b). Then, in 1960, the American paleobotanist Charles Beck described fossil specimens showing the wood and foliage attached to each other. He reconstructed the *Archaeopteris* plant as a forest tree some 20 m tall (Figure 4.17c), and so the extinct class of progymnosperms was born.

It seems unlikely that the *Archaeopteris* tree would have retained the same leaves throughout its life: they were probably replaced many times, so enhancing the flux of carbon from the atmosphere to the soil. Moreover, the wood of *Callixylon* is made up of significant amounts of the complex organic polymer **lignin**, which decays slowly compared with most non-woody plant tissues. Because of the lignin, woody plants have a higher probability of being represented in the fossil record than non-woody (**herbaceous**) plants, and a higher probability of contributing to long-term carbon sequestering.

The progymnosperms are significant because they illustrate clearly the innovations in land-plant biology that brought about significantly increased rates of carbon sequestering from the atmosphere. This effect may indeed have led on to global cooling in the late Carboniferous and early Permian, a topic that we shall explore in greater detail in Chapter 6.

4.7 Green reflections

So far we have considered the effects that the invasion of the land by green plants had on the hydrological and carbon cycles. However, the arrival of plants would also have changed the colour of the land surface from, say, desert orange to forest green. This would have affected the extent to which the Earth reflected incoming solar radiation, and hence the mean global temperature might have been altered.

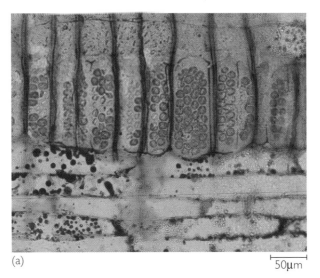

(a)

50µm

Figure 4.17
(a) *Callixylon*, showing general characteristics of secondary xylem. (b) *Archaeopteris* foliage. (c) Reconstruction of the *Archaeopteris* tree.

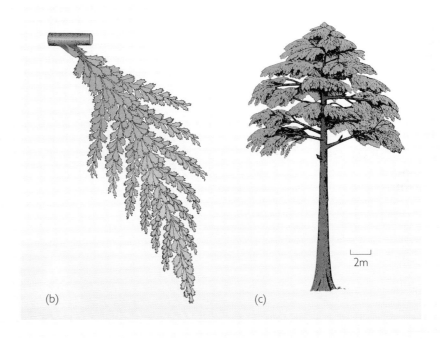

(b) (c)

2m

Let us suppose for a moment that changing the water vapour or carbon dioxide concentrations in the atmosphere had no effect on mean global temperature (which, of course, they must have done), and let us explore how this temperature might have been affected by a change solely in the colour of the land surface.

As you have already seen in *Atmosphere, Earth and Life*, we can estimate the mean global temperature by using what is known as a planetary energy balance equation. In this we are not concerned with the details of the climate, but just with the balance between incoming solar radiation and outgoing long-wave radiation. The outgoing radiation will depend on the surface temperature, T, of the Earth, and that is what we are interested in here.

For our present purposes we can use a modified version of the planetary energy balance equation that was introduced in *Atmosphere, Earth and Life*, and this will make the calculations a little easier:

$$B + CT = \frac{S}{4}(1 - A) \qquad \text{(Equation 4.1)}$$

You will recognize that the right-hand side of the equation is the same as you met before, with S representing the solar radiant flux (here we shall be using a more precise value for today of $1365 \, \text{W m}^{-2}$), and A, the albedo (the proportion of incoming radiation reflected by the Earth). The value of A depends on the distribution of ice, clouds etc., but, globally averaged, it is currently about 0.3 (see *The Dynamic Earth*).

The left-hand side of Equation 4.1 has been simplified in comparison with the planetary energy balance equation that you met earlier; it has also been modified to allow for the greenhouse effect of the Earth's present atmosphere. Most obviously, T is not raised to the fourth power (T^4), as in the earlier version. This simplification applies because a graph of T against the absorbed solar flux (i.e. the right-hand side of the equation) is so steeply exponential that, over the limited range of values that we are interested in, it *approximates* to a steeply inclined straight line of slope C. A further modification arises because, in order to work with the more familiar degrees Celsius for values of T, instead of kelvin, the constant B is put into the equation. This provides a 'false' origin at $0 \,°\text{C}$ for the plot of T versus absorbed solar flux, and also allows for the greenhouse warming effect of the Earth's present atmosphere. You could regard B as the absorbed solar flux that would be necessary to maintain a mean global surface temperature of $0 \,°\text{C}$ if the current composition (and hence the greenhouse effect) of the Earth's atmosphere prevailed. The term C, which is in effect a constant, is there to account for the fact that the global surface temperature may actually be greater than $0 \,°\text{C}$. So, C is the extra amount of absorbed solar flux needed for each $1 \,°\text{C}$ rise in mean surface temperature over the range of values we are interested in. The values of B and C for today's Earth are $B \approx 210 \, \text{W m}^{-2}$ and $C \approx 2.1 \, \text{W m}^{-2} \,°\text{C}^{-1}$.

Substitution of these present-day values of B and C in Equation 4.1 gives:

$$210 \, \text{W m}^{-2} + 2.1 \, \text{W m}^{-2}\,°\text{C}^{-1} \times T = \frac{1365 \, \text{W m}^{-2}}{4} \times (1 - 0.3)$$

which rearranges to:

$$T = \frac{\left(\dfrac{1365}{4} \times 0.7\right) - 210}{2.1} \,°\text{C} = 13.8 \,°\text{C}$$

This value of the calculated mean global temperature T is close to the mean measured global temperature for the present day.

Now we will use Equation 4.1 to calculate the mean global temperature *before* the evolution of land plants. Because we cannot be sure about the amount or type of cloudiness at that time, we shall ignore the possible existence of clouds for the moment and assume that, as now, approximately one-third of the surface of the Earth was covered by land with a desert albedo of about 0.3 and two-thirds by ocean with an average albedo of 0.07. We shall also assume that the solar radiant flux was the same as today.

First, we need to calculate the mean planetary albedo by multiplying the land albedo by its proportion of the Earth's surface (0.33) and adding this to the ocean's contribution (0.07 × 0.67):

mean planetary albedo = (0.3 × 0.33) + (0.07 × 0.67) = 0.146

The mean global temperature would then have been:

$$\frac{\dfrac{1365}{4} \times (1 - 0.146) - 210}{2.1}\, °C = 38.8°C$$

This is an interesting result. It suggests a very warm planet indeed – in fact, one that would have been unrealistically warm.

Question 4.6
Identify two factors that have not been taken into account in the above calculation but that would reduce this mean global temperature.

Next, let us see what effect the colonization of the land by plants would have had on the mean global temperature. As before, for simplicity, we shall ignore cloud albedo. The average albedo for grass (a proxy for primitive land plants) is 0.25, and we shall assume total land surface coverage with such plants. Equation 4.1 becomes:

$$T = \frac{\dfrac{1365}{4} \times (1 - 0.129) - 210}{2.1}\, °C = 41.5°C$$

Question 4.7
Calculate the mean global temperature if the grass were replaced by forest with an albedo of 0.08.

These calculations show the effect of changing the albedo of the land surface, but they also demonstrate an important feedback mechanism. As vegetation developed from grass cover to forest, the reflectivity would have dropped, so more solar radiation would have been absorbed and the planet could have warmed up. Things are not quite as simple as this, however, because increased vegetation would also have increased cloud cover by evapo-transpirational feedback, giving rise to a cooling effect. It is nevertheless worth bearing in mind the effect of vegetation on albedo when considering other feedbacks affecting climate, particularly the greenhouse effect due to CO_2 in the atmosphere, as we shall be doing in the following chapters.

4.8 Summary of Chapter 4

1   Most of the Earth's non-bacterial biomass is in the form of green land plants. The evolution of terrestrial vegetation must therefore have had profound effects on the Earth's surface systems.

2   Before the advent of land vegetation, the continental environment was desert-like and extremely harsh, with large daily variations in temperature. There would have been less cloud cover than at present, higher than current levels of ultraviolet radiation, erratic rainfall, sudden river discharges causing flash flooding, and a mobile mantle of sediment instead of organic-rich soils.

3   Plants adapted to this environment initially by growing close to the substrate surface and so staying within the boundary layer where they were not subject to the more intense desiccation that might occur in the faster-moving free air above.

4   The selective advantages of spore dispersal by wind resulted in vertical growth, which in turn caused the boundary layer to increase in thickness. This is an example of a positive feedback loop.

5   The vertical growth of plants and their consequent greater exposure to desiccation resulted in selection for the following evolutionary innovations:
   ◆ an external cuticle that was largely impervious to water passing through it to the outside air;
   ◆ controllable gas-exchange pores (stomata);
   ◆ a plumbing (vascular) system for the movement of water, nutrients and the products of photosynthesis;
   ◆ structural support tissues;
   ◆ an anchoring system adapted for absorbing water and mineral nutrients from the substrate (usually the ground).

6   Increases in plant height were accompanied by a modified branching system that produced a single main stem with either lateral outgrowths or flattened, webbed, branching systems (leaves) specialized for photosynthesis.

7   Increases in surface area (leaf-area indices) increased the water flux through the air and the carbon drawdown from the atmosphere; both these factors, plus changes in the reflectivity (albedo) of the Earth's surface, affected the global climate significantly.

Chapter 5
A closer look at climate

5.1 Introduction

At the end of the last chapter, we looked at the climatic consequences, in particular, of the complex system of feedbacks between the Earth and its evolving vegetation. In the next three chapters of this book, we shall concentrate on this and other influences upon Phanerozoic climates.

Figure 5.1 represents a qualitative estimate of the changing mean global temperature since the beginning of the Silurian (which roughly coincides with the age of land plants), in relation to that of the present day (marked by a dashed line), which is about 15 °C. There have been only two intervals over the last 443 Ma when the mean global temperature has been consistently lower than that of today. One was in the late Carboniferous and early Permian, and the other more recently, in the Quaternary. These times were characterized by extensive polar ice sheets and we informally refer to the Earth as having been in an '**icehouse**' condition during these intervals. From Chapter 2 you have also seen that there is abundant evidence

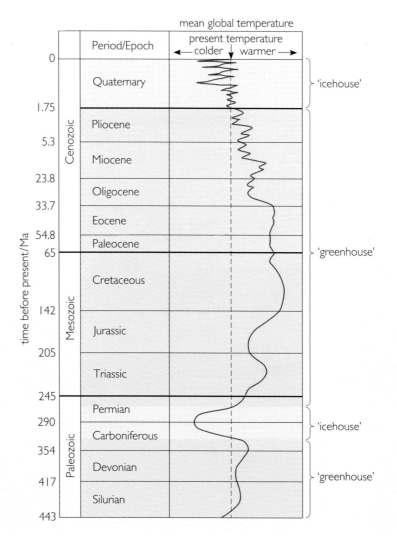

Figure 5.1
Changes in mean global temperature relative to that of the present day (about 15 °C), since the beginning of the Silurian, based on an array of geological evidence. (Note that the time-scale is not linear.)

for icehouse episodes in the late Proterozoic, and further evidence points to such conditions around 2300 Ma and perhaps 2600 Ma ago. For the rest of the time the mean temperature of the Earth was higher than at present – i.e. the Earth was in a '**greenhouse**' condition.

▪ What proportion of the last 443 Ma do these so-called greenhouse conditions represent? (Note that the time-scale in Figure 5.1 is irregular.)

▪ You can see from Figure 5.1 that they represent over 85% of the last 443 Ma.

There is consequently a vast store of information in the geological record relating to how the biosphere was affected by, and perhaps influenced, greenhouse conditions. In other words, the rock record can provide important data relevant to the whole debate about future possible long-term climate change. The Quaternary glaciations, which have shaped our modern world, and which are discussed later in the Course, can be regarded as atypical when we take this longer-term perspective.

We begin, in this chapter, with a brief review of modelling ancient climates, before moving on to two contrasting case studies in the following two chapters. Chapter 6 dwells on the icehouse world of the late Carboniferous and early Permian, together with the subsequent warming that preceded the great mass extinction in the late Permian, while Chapter 7 explores the greenhouse world of the Cretaceous.

5.2 Climate today

5.2.1 How climate works

Climate is, in a crude sense, the integration of weather over time. It includes seasonal variations and is more predictable (or, more correctly in this context, retrodictable) than weather, provided that the factors controlling climate have remained constant over the period being considered. You have already been introduced, in *The Dynamic Earth*, to the basic principles of how climate works, but it is worth briefly revising them here, before launching into the details of modelling past climates.

Atmospheric circulation is driven by heat from the Sun and so any factors that influence the solar energy flux arriving at the Earth will affect climate. An understanding of how global climate works depends on an appreciation of these factors and their effects on atmospheric circulation.

Figure 5.2
Zonal circulation on a smooth (e.g. water-covered) rotating sphere. Latitudinal variation in solar heating causes convection currents (e.g. the Hadley cells), while the influence of the Earth's rotation (the Coriolis effect) causes longitudinal deviations of the air flow (represented by the arrows on the surface of the globe).

5.2.2 Zonal circulation and monsoons

On a perfectly smooth (e.g. water-covered) sphere, barometric pressure contours (and thus many other aspects of climate such as wind patterns, precipitation, etc.) would be parallel to latitude (Figure 5.2). This is commonly called **zonal circulation**. Low pressure occurs at the Equator where solar heating is greatest, and the hot air rises, and high pressure occurs at the poles where cool air descends. The rising air at the Equator draws in air from both hemispheres and this zone of low pressure is therefore called the **Intertropical Convergence Zone** or **ITCZ**. However, low pressure also occurs at 55–60° N and S (with the associated polar fronts), and high pressure at 30–35° N and S (the subtropical highs).

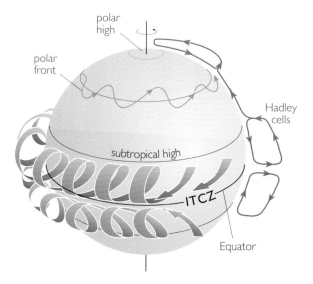

Question 5.1

Can you recall from *The Dynamic Earth* why these intermediate high- and low-pressure belts occur?

Precipitation generally follows the barometric pressure patterns at the surface of the Earth, with higher rainfall occurring in the low-pressure belts and low rainfall in the high-pressure belts.

Question 5.2

Why is high rainfall associated with areas of low pressure?

In high-pressure belts, the converse occurs and the air becomes relatively dry as it descends and warms up through compression.

This zonal picture is generally true over the oceans but is disrupted by a number of factors over the continents. In particular, the thermal contrast between the oceans and land disrupts the zonal pattern. This occurs most strongly in the mid-latitudes where the thermal contrast is greatest. At high mid-latitudes (e.g. around the UK) the ocean tends to be warmer than the surrounding land, particularly during the winter, and the low pressure is intensified over the ocean into low-pressure cells. In low mid-latitudes (e.g. southern Spain), by contrast, the ocean tends to be cooler than the land, and the high pressure is intensified over the ocean. Low- and high-pressure cells may also form over the continents, depending on the degree of thermal contrast within the landmass.

As well as the distribution of land and sea disrupting zonal circulation, major disruptions also occur due to high plateaux. A high plateau, such as that of Tibet, acts as a high-altitude heat source, because the land surface here is, on average, at a higher temperature than the surrounding atmosphere at that height (*The Dynamic Earth*). The effect of the Tibetan Plateau is so strong that it completely overwhelms the zonal pattern in that region. In the summer, a low-pressure cell develops over the plateau due to rising hot air at a latitude where high pressure would be expected in a purely zonal model. The reverse is true in the winter.

Although we associate the word **monsoon** with intense seasonal rainfall over Asia during the summer, the term actually refers to circulation that produces rainfall. All continents, except Antarctica, have monsoonal circulation to some degree because the lower heat capacity of land means that continents tend to warm up faster, and cool faster, than the surrounding oceans. So climates in the real world are a combination of zonal circulation on a global scale modulated by monsoonal circulation on a regional scale.

5.3 Making models of ancient climates

5.3.1 Qualitative models of past climates

Using these basic principles governing climate, it is possible to model, in a qualitative way, the climate for any given geographical configuration in the past. The success of such modelling depends, to a large extent, on the accuracy of the reconstructed geography because, as you saw above, the zonal circulation pattern is disrupted by land/sea contrasts.

Continent/ocean reconstructions based on remanent magnetism (Section 2.2.1) have become commonplace since the mid- to late sixties, following the revolution in the Earth Sciences brought about by the theory of plate tectonics. The reconstructions take into account, not only the continental fits and paleomagnetic data, but also the correlation of rock units, zones of crustal deformation and biogeography (animal and plant distributions). The reconstructed geography can then serve as the framework for a qualitative climate model constructed according to the principles described in Section 5.2.2.

Comparison with the distributions of climatically-sensitive deposits such as coals and evaporites, provides a means of testing the distribution of high- and low-pressure cells and wind patterns predicted by the climate models.

Qualitative models have certain limitations, however, the most important of which is that they cannot be used to predict temperature distributions. For that we must turn to quantitative models.

5.3.2 Quantitative climate modelling

You have already encountered an example of a quantitative model – the application of the simplified planetary energy balance equation discussed in Section 4.7. Useful though these models are, they do not provide any impression of the spatial variation of climatic features or how they might change over short time-scales – hours to decades. To achieve this we need to use far more complex models that are designed to mimic the behaviour of the atmosphere.

Numerical climate models, better known as **atmospheric general circulation models (AGCMs)** work in the following way. The Earth's surface is divided up using a grid system. Above each grid element the atmosphere is divided into a number of layers (up to 19 in the model we will be considering here). In this way the atmosphere is dissected into an array of contiguous boxes (Figure 5.3). The boundary conditions are then set. These boundary conditions include a specification of the physical characteristics of the Earth's surface – whether it is land or sea, its albedo, the rate at which it heats up or cools down, its wetness, and the input of solar radiation. The term 'boundary conditions' refers to all data provided to the model as distinct from model-generated results.

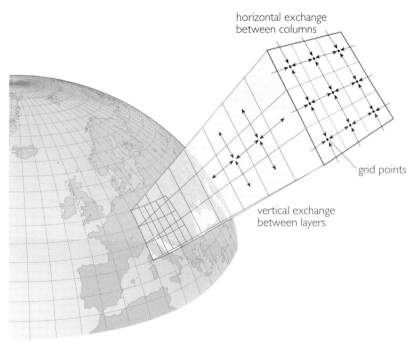

Figure 5.3
Schematic representation of the basic structure of an atmospheric general circulation model – a grid of interacting boxes spread over the globe. (Often the grid squares average 3–4 latitude degrees by 3–4 longitude degrees.) Note that changes in any given box affect those adjacent to it.

The mathematical formulation of the model includes a set of time-dependent equations which describe the interdependent processes taking place in the atmosphere – such as the transport of heat and water vapour by atmospheric winds. The climate is then modelled by integrating these effects box by box.

To describe these processes mathematically for all atmospheric 'boxes' over the whole Earth during a sufficient time to be able to determine the 'average weather' (climate), is far from a trivial task and yet this is what is carried out by the AGCMs. A cluster of supercomputers is normally used and the output is further processed to provide maps such as those shown in Figures 5.4 and 5.5.

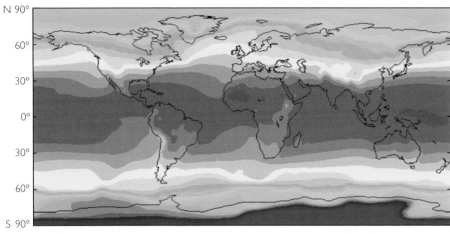

(a) Present day

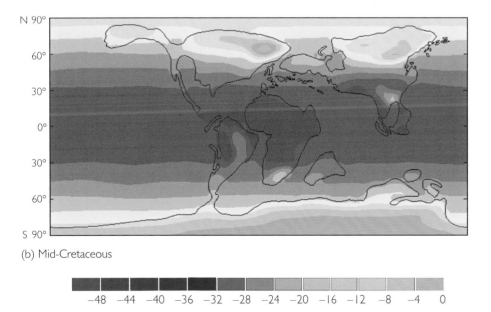

(b) Mid-Cretaceous

mean annual temperature/°C

Figure 5.4
AGCM results showing the mean annual temperature as measured 2 m above the surface (land or sea) for (a) the present day and (b) the mid-Cretaceous.

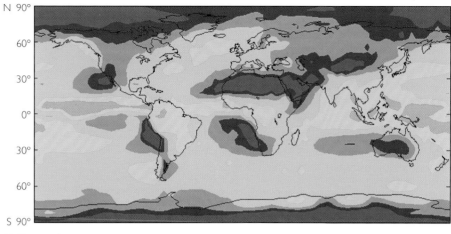

(a) Present day

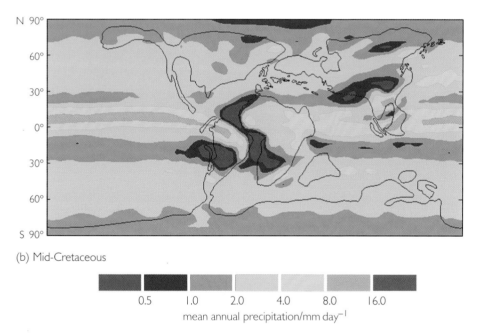

(b) Mid-Cretaceous

0.5	1.0	2.0	4.0	8.0	16.0	

mean annual precipitation/mm day^{-1}

Figure 5.5
AGCM results showing the mean annual precipitation for (a) the present day and
(b) the mid-Cretaceous.

▨   What are the main differences between the present day and mid-Cretaceous
maps in Figures 5.4 and 5.5?

▨   In Figure 5.4, the temperature maps, there is a much stronger Equator-to-pole
temperature gradient in the present world than in that of the mid-Cretaceous.
The mid-Cretaceous is a warmer world overall because of predicted warmer
polar temperatures. Note also the two cool patches in the mid-Cretaceous
simulation – one over what is Greenland today and one over central Russia.
The Greenland cool area is present because the model was told this was an
uplifted area. In the case of Russia the model predicted a very cold winter
and a warm summer, showing the same kind of effect we see today in
continental interiors far from the sea. Unfortunately the latter prediction is
not matched by the geological data from that region which suggest a more

equable climate and a warmer mean annual temperature. At the time of writing this disparity is not understood.

In Figure 5.5, the precipitation maps, we see that the polar regions are predicted to be dry in the present day because the cold poles produce a strong polar high-pressure cell (descending air is dry). This matches reality. In contrast the warmer poles of the mid-Cretaceous are predicted as being wetter because the polar high is much weaker, and at times possibly non-existent. Note also that the mid-Cretaceous equatorial thermal belt is strongly zonal over the larger Pacific Ocean.

As their name implies, these models are really only concerned with the dynamics of the *atmosphere*. They do not explicitly model ocean circulation or the heat transfer that takes place as a result of ocean currents. Some models merely regard the ocean as a 'wet carpet', affording it not only no surface motion but no real depth either. More sophisticated models treat the oceans as if they had a depth of about 50 m as this is effectively the thermally-active depth – i.e. the depth to which the ocean is affected by short-term temperature changes in the atmosphere above. However, even in these more sophisticated models, heat transfer through ocean currents is not simulated dynamically (i.e. the oceans are regarded as being static) but average values of heat transfer derived from other measurements can be supplied to the AGCMs.

Because AGCMs are dependent on thermal characteristics that have to be supplied, their use in reconstructing past climates has to be viewed with caution. We can, of course, supply some empirical data from oxygen isotope studies of the fossilized remains of marine surface living organisms (indicating the temperatures at which their shells formed), but we cannot as yet simulate a fully dynamic system with coupled oceans and atmosphere. The limitation is simply one of computing power. The situation is changing rapidly, however, and within a few years from the time of writing (1996) high-resolution coupled ocean/atmosphere models are likely to be routinely in use.

Given the limitations outlined above, is it worth pursuing the use of AGCMs in the reconstruction of past climates? The answer has to be 'yes', because as we shall see, the models and available geological data are often in agreement, at least to a first approximation. The predictions the models provide are invaluable for investigating how sensitive the climate system is to perturbation and for pinpointing where to look for geological evidence to test the sensitivity or effectiveness of the models. Imperfect though they may be, models of this sort are being used to investigate the rate and pattern of future possible climate change and they are therefore of considerable economic and political significance.

5.4 Summary of Chapter 5

1   The periods in the Earth's history when the mean global temperature was colder than now are designated 'icehouse' times, and periods which were warmer than now are regarded as 'greenhouse' times.

2   Atmospheric circulation is driven by heat from the Sun. Zonal circulation is what would occur if the Earth were perfectly smooth and there were no land/sea thermal contrast. Zonal circulation is only well developed in the real world over large areas of ocean.

3   Monsoonal circulation tends to be seasonal and is driven by strong thermal contrasts. From simple generalizations, it is possible to predict pressure-cell

distributions for any geographic configuration. Ocean waters take longer to heat up and longer to cool down than does land, so in summer low-pressure cells tend to be shifted landward and high-pressure cells oceanward. Similarly a high plateau will be the focus of a low-pressure cell in summer and a high-pressure cell in winter. Such generalizations are the basis of qualitative paleoclimate model predictions.

4   Qualitative models cannot predict temperature regimes: for these, quantitative models are needed. The most sophisticated of these are atmospheric general circulation models (AGCMs) that typically operate on a set of contiguous 'boxes' of atmosphere. The oceans are treated simply; their currents are not modelled. However, this situation is rapidly changing as computers become more powerful. The simplification of the real world means these models may give unreliable results, particularly when dealing with past or future climates when the models are provided with boundary conditions that differ from those of the present.

Chapter 6
An icehouse case study

6.1 Introduction

We look now at the Earth in the late Carboniferous and Permian, an interval spanning some 78 Ma (323 Ma–245 Ma ago). In particular, we shall look at the changes in its geography, climate and life, and proposed explanations linking them. This was an interval in which icehouse conditions gave way to a greenhouse regime, and which ended in the biggest mass extinction recorded in the Earth's history (Section 3.4.2). We focus on two important events, namely the late Carboniferous–early Permian glaciation and subsequent warming (Sections 6.3–6.5), and the late Permian mass extinction (Sections 6.6 and 6.7). The aim of this chapter is not to describe every aspect of life during these times but to concentrate on those organisms which contributed most and/or which were most affected during each event. Thus land vegetation is given prominence in the discussion of the first, glaciation, event (Section 6.4), and then the spotlight is shifted to the marine realm later in the Permian, as we look at the mass extinction event (Section 6.6).

6.2 The setting

6.2.1 Geographical perspective

Students of geology living in Europe (as well as North America) might tend to think of the Carboniferous Period as a time of warm, humid conditions, when the Earth was covered in lush vegetation. Recent history would have been very different without the fortuitous accumulation of Carboniferous plant remains some 300 Ma ago, since these provided the coal reserves upon which the industrial revolutions of the Western World were based. By contrast, the same students might tend to think of the Permian Period as the poor relation – with dry and desolate conditions, on a planet of deserts and salt lakes. However, an Australian or Brazilian student of geology would have a different perspective as the evidence shows that their countries were at times covered by extensive ice sheets during the Carboniferous, and that major coal reserves formed in Australia (as well as South Africa and India) during the Permian.

So the perception of the Carboniferous world, seen from a European or North American perspective as warm and wet, and the Permian as hot and dry is overly influenced by the local geological record. The coal deposits of those continents were produced under tropical conditions, not because the whole planet was warm but because Europe and North America were in low latitudes at that time (*The Dynamic Earth*). In fact, the late Carboniferous and early Permian represent a Phanerozoic cool phase in global terms (Figure 5.1).

To make sense of these varying perspectives, we need only look around our world today to see the incredible variety of climate, environments and life-forms, ranging from tropical rainforests through deserts, temperate forests and tundra to ice-caps. So we might logically expect to find a similar scale of heterogeneity in the Carboniferous and Permian worlds. During these times, however, the continents were arranged very differently from today's (Figure 6.1).

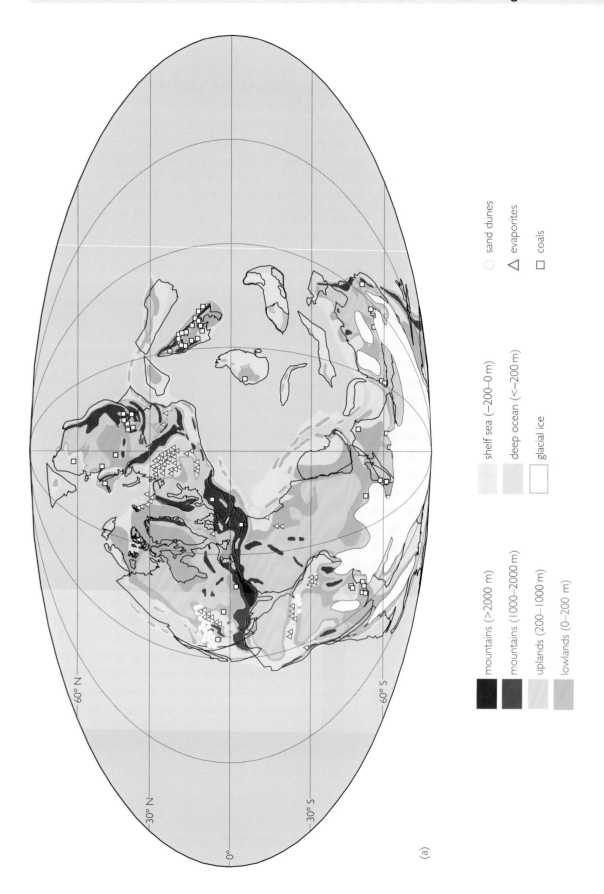

mountains (>2000 m)

mountains (1000–2000 m)

uplands (200–1000 m)

lowlands (0–200 m)

shelf sea (−200–0 m)

deep ocean (<−200 m)

glacial ice

○ sand dunes

△ evaporites

□ coals

(a)

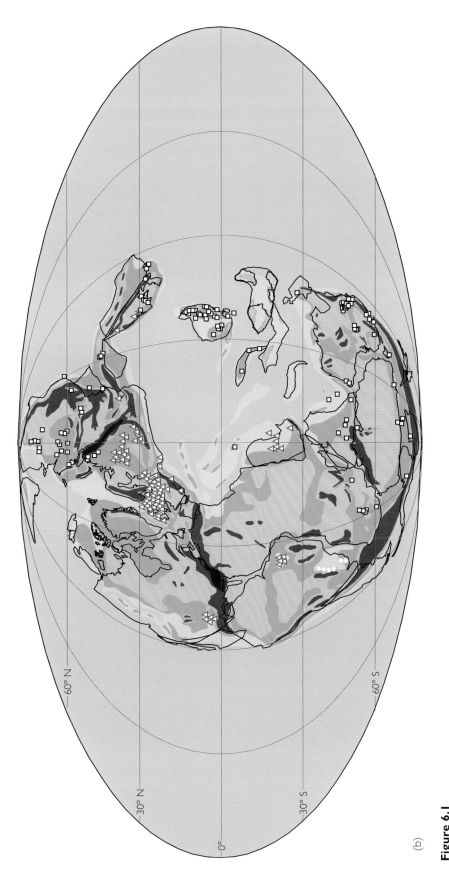

(b)

Figure 6.1
Pangea in (a) the earliest Permian (281 Ma ago) and (b) the latest Permian (247 Ma ago). Climatically-sensitive deposits are also included and are discussed in Section 6.2.2.

Question 6.1
From Figure 6.1 (and what you have read earlier in the Course), what was the most striking aspect of the continental arrangement of these times?

Figure 6.1 shows the position of Pangea in (a) the earliest and (b) the latest Permian, and illustrates the changes in location of the continent and the shift from glaciated to ice-free conditions. These maps, compiled by Fred Ziegler and colleagues at the University of Chicago, represent the most detailed paleogeographical synthesis to date of a vast array of data. Figure 6.1a can serve as a general reference map for the Permo-Carboniferous (i.e. the late Carboniferous to early Permian) icehouse world and Figure 6.1b sets the scene for the second (extinction) event which is discussed in Section 6.6. You will need to refer back to them throughout this chapter, since they provide an important framework for the topics discussed here.

Life on Pangea was very different from that of today: there were no flowering plants and no mammals. As far as the oceans were concerned – recall from Chapter 3 that the Mesozoic Marine Revolution was yet to come – predation on shelly organisms was less intense than that of today, and sessile shelly forms lying exposed upon the sea floor were commonplace. Neither, apparently, was there any calcareous plankton. Yet there are a number of similarities in terms of the broad effects of global climate change, including the response of organisms to it, which might inform our speculations about the effects of climate change in the future.

6.2.2 Geological evidence

In Section 5.3.1, we noted that the distribution patterns of climatically-sensitive deposits, such as evaporites, red beds, glacial diamictites (or 'tillites') and coals can provide important clues to past climates. As you will see below, fossil wood can also help.

The formation of evaporites (salt deposits) requires that evaporation exceeds precipitation. Ideal conditions occur in arid regions, where formation can take place in enclosed basins with high temperatures and low rainfall. Modern evaporites occur mostly in subtropical regions centred around 25° to 35° latitude. A similar distribution can be seen in Figure 6.1 with respect to the inferred geography of the Permian. Much the same pattern can be seen for desert sand dune deposits.

Red beds are sedimentary rocks containing haematite (Fe_2O_3) that formed under oxidizing conditions in a hot climate (*Atmosphere, Earth and Life*). Often the original source of the iron was exposed igneous or metamorphic rock which had been intensely chemically weathered in hot, humid, oxidizing conditions. Later this iron could be remobilized (as Fe^{2+}) in anoxic groundwater and re-precipitated in desert sediments, as iron oxide again, when evaporation drew the water up towards the surface. Modern red beds have formed largely within 30° of the Equator (reflecting Pleistocene shifts between humidity and aridity) and most Paleozoic red beds seem to have had a similar distribution, being commonly associated with evaporite deposits. Of course, such ancient red beds may be eroded to yield red soils at a later date in a different latitude, as can be seen, for example, in the countryside of Cheshire.

In contrast to these indicators of climatic warmth, most diamictites are considered to have been deposited by glaciers, as tillites (Box 2.1). Their widespread occurrence in southern areas of Pangea during the Permo-Carboniferous indicates that large areas experienced glacial conditions at least some of that time.

The formation of coal requires, among other things, a net surplus of precipitation over evaporation, sufficient warmth and light for plants to grow, and isolation of buried plant material (or peat, the precursor to coal) from the oxidizing atmosphere. Rainfall and plant productivity are closely linked, while periods of drought strongly affect preservation because falling groundwater levels permit oxidation of organic matter (through aerobic decomposition or burning).

Further clues come from growth rings in trees which can also serve as climate indicators (Box 6.1). Fossil Carboniferous tree trunks from low paleolatitudes lack, or have only faint evidence of growth rings, indicating that they grew in near-constant conditions of humidity and temperature. Therefore, we can deduce that most Carboniferous coals formed under tropical conditions at low latitudes. By contrast, thick coal deposits also formed in the Permian within 5° to 30° of the South Pole (to be discussed in Section 6.4.3). These coal deposits contain fossilized tree trunks with prominent growth rings, implying seasonal growth.

So, from our knowledge of the present-day distributions of these different climatically-sensitive deposits, and the climate conditions under which they are laid down, we can infer similar conditions when we encounter them in the geological record.

Another important clue to ancient climates is the isotope record. As mentioned in earlier chapters, variations in the strontium isotope ratios of marine limestones can be used to interpret the tectonic activity and climatic conditions that prevailed at the time of their deposition. Similar clues are also available from carbon isotope data.

By now you should have a general picture of the geography and climate of the Carboniferous and Permian, as well as some of the methods geologists have used to interpret them. We will now focus, in detail, on some of the more important aspects of the Earth and its life within the interval spanning the late Carboniferous and early Permian.

Box 6.1 Tree growth rings

Growth rings are formed when a tree grows at varying rates over time. Typically, variations in wood growth are caused by environmental changes in, for example, water availability, temperature or light regime (day length). Anywhere away from low latitudes such variation is usually tied to the annual cycle of seasons. Wood cells with large internal cavities are produced early in the growing season, when the availability of, and the demand for, water is high. These cells comprise the **earlywood**. When water is less available and demand is less, in late summer and autumn, the water-conducting space in the cell becomes constricted as the cell walls thicken. These **latewood** cells are mechanically stronger than earlywood cells, but less efficient at water conduction. No cells are produced during the period of winter 'shutdown' or dormancy of the tree. The change from large cells to small thick-walled ones thus corresponds to seasonal changes in tree growth, forming a single growth ring. In non-seasonal environments, such as in tropical rainforests, there is less of a fluctuation in climate conditions through the year (though there may be some variation in rainfall) and growth is consequently more uniform, with rings being absent or only weakly developed. The ratio of wall thickness to cell cross-sectional area for water conduction is a compromise to meet both water demand and that for structural strength.

6.3 Permo-Carboniferous glaciation and subsequent warming

6.3.1 The available record

As you have seen in Figure 5.1, the late Carboniferous and early Permian interval is interpreted as having been a time of pronounced cooling. This was apparently followed by a shift back towards warmer conditions during the Permian, with global climate eventually becoming warmer than at present.

■   Take another look at Figure 5.1, in particular the time-scale on the left. Why do you think so much space (nearly 50% of the diagram) is devoted to the last 65 Ma, though it represents only about 15% of the time shown? Related to this, why do you think the temperature curve changes from being smooth to jagged the closer we get to the present?

■   The scope and detail of the record is better for younger periods. There has been less time for the younger rocks to be eroded or deformed, and so the information they contain in terms of rock type, fossil content and chemical composition is greater the younger they are. Hence, a relatively large amount of space is devoted in Figure 5.1 to the last 65 Ma, and the jagged nature of its temperature curve reflects the more detailed record that is available. We know less and less the further we go back in geological time and our interpretations tend to become less certain as extrapolation between data points becomes greater, as a reflection of the increasing gaps in the rock record.

The perceived variation in the quality of the record is further exaggerated by research bias among scientists: the excellence of the Quaternary record, in particular, has led to a vast amount of published data.

So how much do we really know about Permo-Carboniferous conditions? The answer is a fair amount. Even though it is often harder to work on rocks of that age than on more recent rocks – a bit like walking in the dark with a light only occasionally being switched on to give an idea of what's around – a reasonable picture has been built up by studying the available evidence. Some of this will be discussed in the following sections.

6.3.2 The link between atmospheric CO_2, vegetation and climate

We all know only too well that human-induced buildup of atmospheric CO_2 (and other greenhouse gases) is implicated in rapid global warming. On a longer time-scale than that affected by our interference, atmospheric CO_2 levels appear also to have varied throughout the Earth's history, as was demonstrated in *Atmosphere, Earth and Life*. We can surmise that such variations must have affected global climate, through the potent greenhouse effect of CO_2.

Some of the many and complex ways in which atmospheric CO_2 levels rise and fall were discussed in broad terms, in the context of the GEOCARB model, in *Atmosphere, Earth and Life*. We will restrict ourselves in this chapter to those which are most relevant to the time interval we are considering. Two major routes via which CO_2 can be removed from the atmosphere are through direct uptake by plants (during photosynthesis), and by dissolution in water. So long as the rate of photosynthesis exceeds that of respiration, as a consequence of the burial of organic material, the net effect, in the longer term, is sequestration of atmospheric

CO_2. On land, the buried organic material first becomes peat, then coal, so that a reservoir of organic carbon builds up, locked away from the atmosphere.

Carbon dioxide is also highly soluble in water and this provides a second major route by which it is removed from the atmosphere. The overall reaction between carbon dioxide and water to form a weak acid, carbonic acid (H_2CO_3), can be expressed as:

$$CO_2 + H_2O \rightleftharpoons H_2CO_3 \qquad \text{(Equation 6.1)}$$

As you have seen earlier in the Course, carbonic acid can dissociate to release a hydrogen ion (H^+) and the bicarbonate ion (HCO_3^-), which in turn dissociates to form the carbonate ion (CO_3^{2-}) and a further hydrogen ion. The high levels of CO_2 in vegetated soils (from respiration of plant root systems, and microbial decomposition) means that CO_2 concentrations are typically 10 to 100 times higher in the soils than in the atmosphere. It is this acidic water which is mainly responsible for the weathering of minerals in soil and rock.

Question 6.2
What are the two main rock weathering reactions, involving dissolved CO_2, that were considered in *The Dynamic Earth* and *Atmosphere, Earth and Life*, and which of these is considered to lead to a net drawdown of CO_2 from the atmosphere when the ensuing precipitation of carbonate is allowed for?

We also need to consider a third weathering reaction – that involving exposed deposits of peat and coal. In contrast to the weathering reactions just discussed, this oxidative process *releases* CO_2, while, of course, using up oxygen. (You might think of this process as a very long-term equivalent of the respiration equation in Section 4.5.)

Therefore, plants are responsible, both directly and indirectly, for considerable drawdown of atmospheric CO_2. The Permo-Carboniferous interval seems to have been outstanding in terms of the direct effect of vegetation on climate. Two significant things occurred. The spread of land plants, particularly trees, provided a large pool of organic matter, some of which was then buried. In addition, since the Devonian (which you read about in Chapter 4) they had greatly increased the rate of soil and rock weathering. The result of the weathering process was an increased transfer of CO_2 from the atmosphere to the oceans, in the form of bicarbonate ions (see the answer to Question 6.2). Ultimate burial there as limestone meant that atmospheric CO_2 levels were lowered. Although CO_2 is returned to the atmosphere through volcanism and weathering of exposed organic deposits, it seems that this reverse process was greatly outweighed during the late Carboniferous and early Permian by CO_2 drawdown mediated, in one way or another, by land plants. Models for the change in atmospheric CO_2 levels (such as the GEOCARB model), based on the estimated balance of carbon fluxes, show a marked fall in CO_2 for this interval (Figure 6.2: repeated from *Atmosphere, Earth and Life*). This in turn would have resulted in a lowering of mean global temperatures.

However, other mechanisms can also be invoked to explain the globally cool conditions during this time interval, in particular the location of the southern part of the Pangean landmass over the South Pole, and the effect that the assembly of such a supercontinent might have had on climate. For this we need to take a look at the influence of plate tectonics.

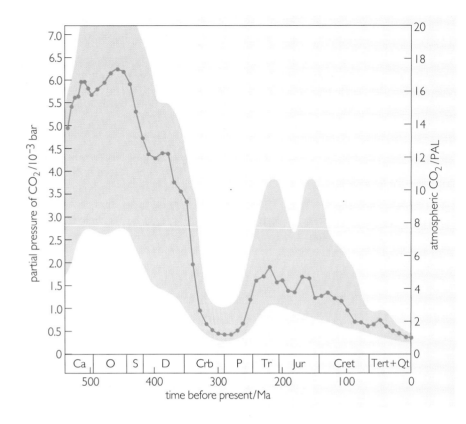

Figure 6.2
Variation in the level of atmospheric CO_2 over the Phanerozoic as calculated from the GEOCARB model. Note the pronounced fall in CO_2 levels around the Permo-Carboniferous boundary, to levels similar to those of today.

6.3.3 Continental motion, collisions and climate

The processes of plate tectonics were explained in *The Dynamic Earth*. The following aspects are relevant to the Permo-Carboniferous icehouse world:

◆ the configuration and position of land with respect to the poles, and their effect on ocean circulation;

◆ volcanic activity related to mid-ocean ridge and subduction processes;

◆ collisional processes causing uplift and mountain belts;

◆ eustatic changes in sea-level.

As you shall see in the following discussion, each of these can affect climate, both locally and globally.

As noted above, the cause of the Permo-Carboniferous glaciation has, in some quarters, been put down to the positioning of a part of the Pangean landmass at that time over the South Pole.

Question 6.3
Which of the hypothetical patterns of continent distribution discussed in *The Dynamic Earth* ('ring', 'cap' or 'slice' world) fits with this suggestion?

As you saw in *The Dynamic Earth*, a 'cap world' configuration is expected to lead to a markedly cooler (south) polar region. If snow precipitation exceeded ablation (in areas of maritime influence, with moist air coming off the ocean), its accumulation could have led to the formation of an ice-cap. However, a counter-argument is that landmasses were positioned over the poles at other times in the Earth's history, including much of the Cretaceous, when global climate was warmer than today. So the onset of glaciation and subsequent melting of the polar ice-cap cannot be explained solely by movement of the Pangean landmass on to, and away from, the South Pole.

An arrangement more reminiscent of a 'slice world' was reached in the Triassic, when the landmass was symmetrical about the Equator. This had developed gradually over millions of years, through the northward drift of Pangea. Western equatorial regions of Pangea had thus started to dry out first (as expected from the 'slice world' model described in *The Dynamic Earth*) at the very end of the Carboniferous, and arid conditions then spread eastwards as the monsoonal system developed and further disrupted zonal circulation. This can be seen in the rock record, with a change from indicators of wetter conditions, such as coals, to those indicative of aridity (e.g. evaporites and red beds). Thus, conditions became progressively drier at low latitudes during the Permian, and the development of a monsoonal system probably contributed to this. However, these changes do not explain the overall deglaciation and global warming.

If there was only one Pangean supercontinent, then it follows that there was only one 'superocean'. However, there were still some Asian microcontinents drifting around (Figure 6.1), some presumably with small spreading ridges between them. Nonetheless, compared to times of continental fragmentation, the total length of mid-ocean ridge systems, and corresponding subduction zones, would have been less than at other times, so delivering less CO_2 to the atmosphere. Though this may help to explain the glaciation (through net drawdown of CO_2, as discussed earlier) it will not help us to explain the subsequent retreat of glaciers during the Permian.

We have a third tectonic factor to consider in the Pangean supercontinent and its effect on global climate – the collision of all the plates which contributed to its formation, and the consequent mountain chains they created. In *The Dynamic Earth* you read about the uplift of the Himalayas and how they formed through the collision of India with mainland Asia. The assembly of Pangea from its previously wandering parts was akin to many such continent–continent collisions. Therefore numerous mountain ranges were built, though whether anything like the Tibetan Plateau formed is open to conjecture.

Question 6.4
What effect might mountain building on such a scale have had on the rate of CO_2 drawdown from the atmosphere?

Without compensation for any such increased drawdown, the consequence would have been global cooling. It is known that major mountain chains, generated by such collisions, were around in the late Carboniferous and early Permian. The eroded relics of these still form many upland areas in northwestern Europe (e.g. the Ardennes and Harz Mountains) and the Appalachian Mountains in the United States, for example, but how can we evaluate the effect of their formation on CO_2 levels and climate?

■   Can you recall from Chapter 2 the two major influences, and their effects, on the ratio of the two strontium isotopes ^{87}Sr and ^{86}Sr in seawater?

■   High $^{87}Sr/^{86}Sr$ ratios occur in the weathering products of continental rock (including mountains), whereas low ratios occur in the hydrothermal effusions associated with sea-floor spreading. Since all of this material ends up in the oceans, the values preserved in marine carbonate rocks, in which strontium is incorporated, reflect the relative inputs of continental weathering and sea-floor spreading through geological time.

Now consider the likely pattern in the $^{87}Sr/^{86}Sr$ ratio in seawater for the interval in question.

Question 6.5

Which of the two influences on the strontium isotope ratio was likely to have been more important in the Permo-Carboniferous, and would $^{87}Sr/^{86}Sr$ ratios have been relatively high or low?

Figure 6.3 shows a compilation of strontium isotope ratios throughout the Phanerozoic. In the long term, many factors are likely to have been involved in the overall pattern. Focus your attention, for the moment, on the Carboniferous record, approaching the late Carboniferous-early Permian glaciation.

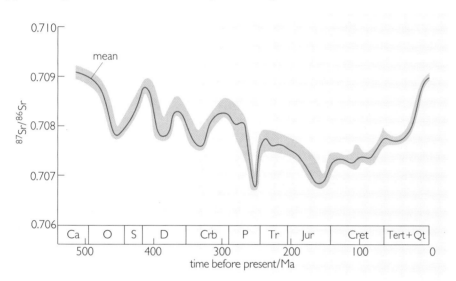

Figure 6.3
$^{87}Sr/^{86}Sr$ ratios in marine limestones through Phanerozoic time.

▨   Is the prediction from Question 6.5 borne out by the evidence given in Figure 6.3?

▨   Yes. Note the relatively high $^{87}Sr/^{86}Sr$ ratios for the interval in question, particularly with respect to the early Carboniferous (before the main continental collisions), reflecting increased continental weathering relative to sea-floor spreading.

Thus mountain building may also help to explain the globally cool conditions, through increased drawdown of CO_2 during weathering of exposed rocks. The relationship is not quite so simple in detail, however, as continental weathering rates also vary with temperature (as do most chemical reactions). Hence you might expect increased continental weathering rates (yielding high $^{87}Sr/^{86}Sr$ ratios) to be associated with *high* atmospheric CO_2 levels and global warming. The drawdown of CO_2 could then regulate atmospheric CO_2 levels, and hence climate. The converse compensation should occur with lowered CO_2 levels leading to global cooling, with reduced weathering rates. But, as you have just seen, *increased* weathering rates have been inferred during a time of globally cool conditions. So how do we explain this apparent paradox? It seems that the sheer volume of rock made available (through major uplift) for weathering was the over-riding factor. Also, there were major collisional belts in low latitudes, where weathering rates would have been high. We know from the Himalayas today that much of the chemical weathering of the eroded sediment may eventually take place in lowland areas flanking the mountains. This later processing, combined with the large-scale effects of vegetation, could have resulted in abnormally high continental weathering rates in the late Carboniferous to early Permian, global cooling notwithstanding, so further enhancing the icehouse conditions.

Finally, tectonic processes can cause global sea-levels to change, which may in turn affect global climate. Sea-level variations can occur through changes in the volume of the ocean basins. As you have read earlier in the Course, this is caused, most directly, by changes in the volume of mid-ocean ridges.

Question 6.6
How would you expect the plate tectonic activity of the time to have influenced global sea-levels?

Low sea-levels would have resulted in the emergence of vast areas of former continental shelf. Lowland swamp forests would have expanded, with a consequent rise in sediment weathering rates and drawdown of CO_2 through peat accumulation. Global sea-levels would also have been lowered yet further due to the accumulation of polar ice, locking up large volumes of water. Although this does not explain the initial formation of the polar ice-cap, it may have been a factor in maintaining icehouse conditions.

So there are several tectonic factors, each of which may help to explain the conditions seen in the Permo-Carboniferous icehouse world. They were also often intertwined with the effects of vegetation and CO_2 levels on climate. We will focus on the effect of vegetation in Section 6.4, but for now let us briefly take an external view of our planet.

6.3.4 Extraterrestrial causes of climate change

There is another factor involved in climate change which we need to consider, namely the increase in solar radiation throughout the Earth's history. The solar flux during the Paleozoic has been estimated as being 3–5% lower than that of today. This might not seem much, but an increase in solar radiation of only 2% has an effect on mean global temperature which is equivalent to a doubling of atmospheric CO_2 levels. Therefore 3–5% less solar radiation should account for a considerable lowering of mean global temperatures, relative to today. As Figure 6.4 shows, the combination of low atmospheric CO_2 levels (derived from the GEOCARB model in Figure 6.2) and reduced solar radiation results in an even sharper contrast between the Permo-Carboniferous icehouse world and other geological intervals (over the Phanerozoic).

So, lower levels of solar radiation can also be invoked to explain, in part, the Permo-Carboniferous climate.

Question 6.7
If the level of solar radiation has increased over time, why wasn't the pre-Carboniferous climate even colder?

However, if solar radiation has increased only gradually through time, then it cannot fully explain both the marked global cooling of the late Carboniferous and early Permian *and* the relatively rapid subsequent warming during the Permian.

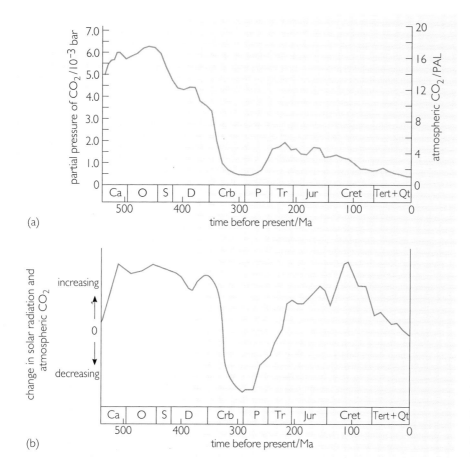

(a)

(b)

Figure 6.4
(a) Phanerozoic atmospheric CO_2 levels derived from the GEOCARB model shown in Figure 6.2. (b) Combined plot showing net forcing effect of CO_2 and increasing solar radiation on climate through the Phanerozoic, relative to the present level. Note the significantly lower levels around the Permo-Carboniferous boundary compared to those of today.

6.4 The impact of land vegetation

6.4.1 Land plants in the Carboniferous and Permian

Given the striking effect that Permo-Carboniferous land vegetation appears to have had on CO_2 levels and global climate, we should look at it now in more detail. You learned in Section 4.6.4 that most major plant groups had evolved by the end of the Devonian. However, things really took off in the Carboniferous, with much diversification and the appearance of extensive low-latitude forests dominated by plants a bit like very large versions of living clubmosses (lycophytes) and horsetails (sphenophytes). Many of these grew as tall trees dominating the low-latitude swamps.

▨ Why can height be advantageous to a plant?

▧ Height enables a plant to compete successfully for access to light and to disperse its spores by winds over a wider area (Section 4.6.2). Consequently, a tall plant would have a distinct advantage over its shorter competitors.

There was a plentiful supply of water at these low latitudes and many plants evolved and thrived. As a result, thick peat deposits formed in the late Carboniferous low-lying wetland areas, eventually producing the major coal reserves exploited by us today.

Later, towards the end of the Carboniferous, the lycophyte trees dominated in only a few areas, with seed plants such as conifers becoming dominant in a number of habitats. Seeds are able to survive periods of drought, germinating when external conditions become favourable: with a seed, an embryo is shielded and nourished by nutritive tissues within a protective coating. These changes in vegetation signalled the gradual loss of most low-latitude peat-forming swamps and their widespread replacement by drier, well-drained environments in the early Permian.

We will look at 'post-glacial' Permian vegetation a little later. For now, let us concentrate on the life and death of the agents of CO_2 sequestering in the icehouse world – the low-latitude swamp forests of Europe and North America.

6.4.2 *Lepidodendron* – a typical swamp dweller

Rather than comprehensively surveying the kinds of trees that made up the late Carboniferous swamp forests, we will concentrate on the features of one in particular – *Lepidodendron*, a typical example of a lycophyte tree. It is one of the most thoroughly studied of all Carboniferous plants. Not surprisingly, it figures prominently in Carboniferous swamp forest reconstructions (Figure 6.5), and you can even see splendid examples of the real thing in the Fossil Grove in Glasgow.

(a)

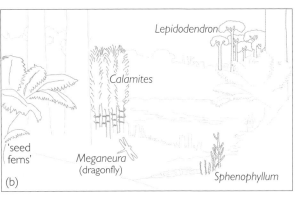

(b)

Figure 6.5
Reconstruction of a late Carboniferous low-latitude swamp forest. Major plant types are shown in the key.

Lepidodendron was the dominant forest canopy form during the Carboniferous and grew to 30 to 40 metres in height when mature. The tree only developed a branched crown when in its reproductive phase. In spite of its large size, it was 'cheaply' constructed in that it had very little woody tissue. Based on this, it is thought that one could have chopped through a 1-metre trunk with nothing more than a machete! Instead of wood, the trunks were largely supported by bark, which means that each tree may have lived for only one or at most a few seasons before falling over and perhaps being buried. The vigorous growth and short lifespan of these plants helps to explain why so much peat accumulated in these swamps, eventually to become coal. Peat accumulates if the rate of production exceeds that of decay. Studies of modern trees show that bark contains chemicals that make it more resistant to biological decay and can also inhibit decay of other substances, such as carbohydrates and forest litter. Consequently, there is little sign of decay in late Carboniferous plants and the short-lived *Lepidodendron* trees could 'rest in peace' as they were buried in the swamps. Some examples of their fossilized remains are shown in Figure 6.6.

Near the base of the *Lepidodendron* trunk, the primary water-conducting tissue (**xylem**) was restricted to a cylinder only a few centimetres in diameter. Although there wasn't very much of it, the xylem must have been able to conduct water and nutrients from the roots up to the top of the tree. However, it seems there was no, or extremely little, **phloem** present in a *Lepidodendron* trunk. Phloem is the vascular tissue which enables movement of the products of leaf photosynthesis (such as sugars) to the rest of the plant, including the roots. So how did the root systems of these trees get their food?

The rooting structure of the *Lepidodendron* tree (their fossils are known as *Stigmaria*) terminated in numerous finger-width hollow rootlets which were helically arranged on root branches (Figure 6.6a and c). In many respects these projections appear to be more similar to leaves than roots, but leaves which have been modified for anchorage. The root systems were shallow and some of the rootlets could have been exposed to sunlight filtering through water in the swamp. So the three vital ingredients for photosynthesis – H_2O, CO_2 and energy – would all have been available in the shallow swamp waters. It is therefore possible that the roots and leaves photosynthesized and nourished themselves independently.

6.4.3 The legacy of the forests

Plants like *Lepidodendron*, which were superbly adapted to, and dominated, low-latitude swamp environments, were important because they enhanced CO_2 drawdown from the atmosphere. But why did they become extinct?

These lycophytes, which are only remotely related to seed plants, had, in fact, developed a reproductive structure analogous to a seed. Indeed, differences between a lycophyte reproductive structure and a 'true' seed are largely technical. However, if you think back to the advantages mentioned earlier of seed plants and their ability to survive and even thrive in drier habitats, there seems to be a contradiction: why did the lycophytes such as *Lepidodendron* die out when the swamps dried up? As we have seen, they had highly specialized rooting organs which spread out in shallow sediments and were well adapted to support a tree living in waterlogged habitats. The overly specialized adaptations of their underground organs was the most likely cause of their extinction, rather than the failure of their reproductive systems. There are probably two reasons for this: growth in drier and firmer soils would seem to have been very difficult, if not

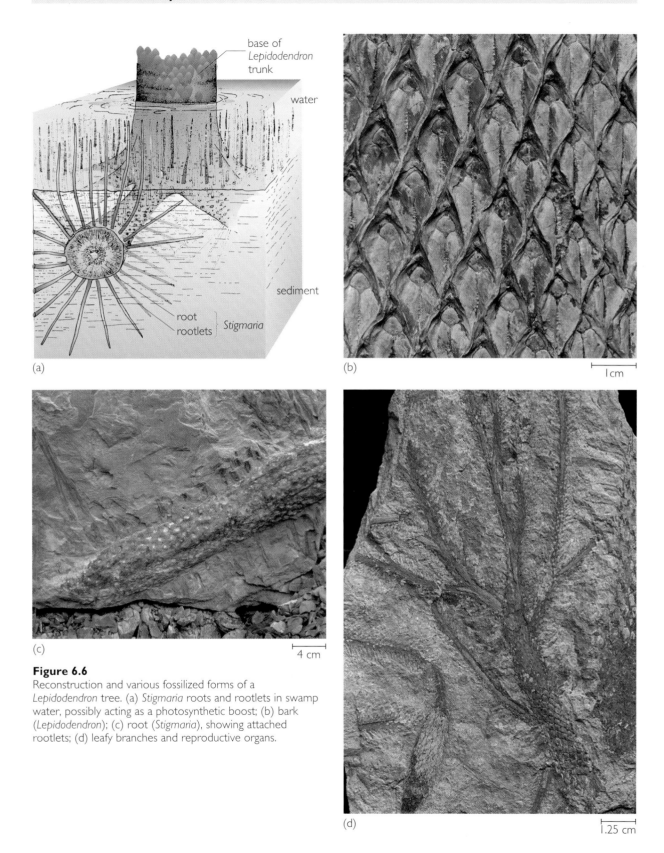

Figure 6.6
Reconstruction and various fossilized forms of a *Lepidodendron* tree. (a) *Stigmaria* roots and rootlets in swamp water, possibly acting as a photosynthetic boost; (b) bark (*Lepidodendron*); (c) root (*Stigmaria*), showing attached rootlets; (d) leafy branches and reproductive organs.

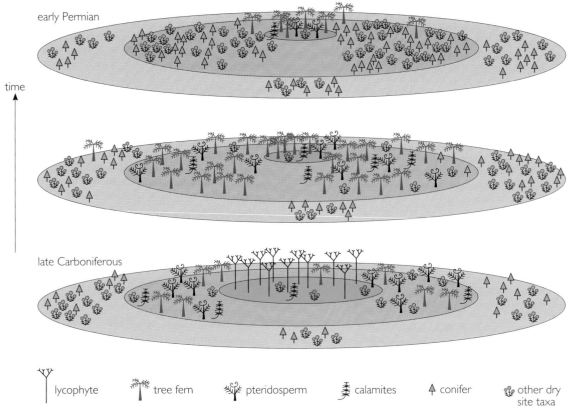

early Permian

time

late Carboniferous

lycophyte   tree fern   pteridosperm   calamites   conifer   other dry site taxa

Figure 6.7
Late Carboniferous-early Permian change in equatorial, European/North American plant communities. The dominant lycophytes of the late Carboniferous gave way to previously marginal plants (e.g. conifers) adapted to drier habitats. The inner circle represents the wettest environments, the middle band the non-swamp habitats which were progressively affected by seasonal dryness into the Permian, and the outer band the uplands that were rarely wet.

impossible, for the shallow-rooted *Stigmaria* with its presumably delicate rootlet apices, and, by being buried, the latter would have lost the energy (sunlight) needed to nourish themselves. Thus, with the demise of lycophytes such as *Lepidodendron*, the way became open for the remaining seed plants to dominate and diversify (Figure 6.7).

Some of the seed plants had leaves adapted to life in dry environments and became dominant elements of lowland European and North American vegetation in the early Permian. As you can see in Figure 6.7, a number of these already existed in the late Carboniferous, but were marginalized in drier habitats on the fringes of the wetlands; they simply seized upon the opportunity presented to them as the low-latitude coal forming swamps of Europe and North America gradually disappeared.

It should be stressed that these changes in low-latitude vegetation and climate did not all occur at the same time on a global scale. Further east, in Asia, broad-scale patterns of vegetation and climate remained pretty similar from the Carboniferous right through until the end of the Permian, when conditions finally became more arid. In contrast to Europe and North America, China has major Permian coal reserves.

1 cm

Figure 6.8
Numerous *Glossopteris* leaves preserved in layers of fine-grained sediment.

Question 6.8
Why did these changes in vegetation occur at different times in different places?

Elsewhere, extensive forests had developed in high southern latitudes during the early Permian. These were dominated mainly by the **glossopterids** – deciduous trees with distinctive tongue-shaped leaves. These leaves formed great mats where they were shed and buried annually in the swamps (Figure 6.8). Glossopterids have a particular historical claim to fame in that they played a major role in the original reconstruction of Pangea by Alfred Wegener (among others) early in the 20th century. He noted that they are found in several areas across the present southern continents and so used them to help show the assembly of those continents in the past. The vegetation must have been abundant, since the dead remains accumulated as thick peat deposits (providing commercially-mined coal reserves in Australia, for instance). From what you read earlier about the extensive southern ice sheets, it may seem strange that such thick peat deposits should have accumulated shortly after the ice retreated from this area. In contrast to the warm, humid conditions at low latitudes in the icehouse world, these peats formed in cool, swampy bogs at high latitudes in the newly developing greenhouse world. We will return (in Chapter 7) to the question of how such productivity could have occurred at high latitudes when we look at similar forests that grew some 100 Ma later, in the even more extreme greenhouse world of the Cretaceous.

6.5 A synthesis for the icehouse

Intense glacial conditions typified the late Carboniferous and early Permian, but why did they occur and persist? It is difficult to point to one single factor, though the role of plants seems to have been crucial. The timing of the maximum spread of Carboniferous coal-forming swamps coincided with minimum estimated CO_2 levels and maximum glaciation. However, the impact of land plants was inextricably linked with other factors. The relative reduction of mid-ocean ridge systems would have resulted in a sea-level fall, providing extensive lowlands for the coal-forming swamps. Moreover, the reduced volcanic activity would have produced less CO_2 to counteract the effects of the biotic sequestering. The accumulation of polar ice would itself have ensured further falls in global sea-level, resulting in yet more exposure of continental shelf area. In addition, movement of part of the landmass over the South Pole may also have played a role in the observed cooling. Finally, the formation of Pangea resulted from major collisional processes, which caused continental uplift and subsequently increased weathering rates. The copious sediment shed from these mountains also contributed to the preservation of coal by continually burying the swamp peats in vast delta systems. Together these factors established a sustained net drain on atmospheric CO_2, so reducing temperatures, and maintaining icehouse conditions.

These conditions were followed by deglaciation towards the end of the early Permian, resulting in only a few small ice-caps remaining on the highlands of southern Africa. Cool humid conditions prevailed over the rest of the southern part of the landmass, enabling the growth of glossopterids and the formation of high-latitude coal deposits. There was a subsequent loss of even these remaining ice-caps, with the further development of globally warm conditions in the late Permian. The movement of Pangea away from the South Pole, so as to straddle the Equator in the Triassic, gave rise to the development of an extremely strong monsoonal system, with pressure systems and precipitation zones swinging seasonally from hemisphere to hemisphere. Warming would have been enhanced by exposure and oxidation of some of the organic carbon that had been sequestered during the late Carboniferous and early Permian. Indeed, the loss of most coal-swamp vegetation may have provided another positive feedback to the warming, since there would have been less capacity for CO_2 removal from the atmosphere. It seems that tectonically-driven influences on climate now overwhelmed any cooling and related change due to CO_2 reduction by biotic sequestering. The result was a greenhouse world which lasted for some 250 Ma (Figure 5.1).

6.6 The late Permian mass extinction

6.6.1 Life at sea

So far in this case study the main biological role has been played by land plants. We shall now turn our attention to marine life, which is important for understanding our second 'event' – the late Permian extinction – and the conditions leading up to it. Many of the factors explored as possible causes of icehouse conditions and their termination crop up again here, too, along with a few additional twists. No other recorded extinction event had such a catastrophic effect (Section 3.4.2).

The pattern of extinction at the close of the Permian was by no means simple. It has been estimated that the episode of mass extinction occurred over some 3 (perhaps as many as 8) Ma and so was not an instantaneous single event. Indeed, recent analyses suggest that there may have been two distinct episodes of extinction. Moreover, some groups of marine animals continued as if nothing had happened, others were already in terminal decline anyway, and still others recovered and diversified somewhat later in the Triassic. These biases are reflected in the plot of family diversity shown in Figure 3.8.

> **Question 6.9**
> What were the relative effects of this mass extinction on the 'evolutionary faunas' recognized in Figure 3.8?

On land, by contrast, major changes occurred not only at the end of the Permian but also much earlier in the period, apparently as a result of climate change related to the termination of icehouse conditions and the onset of a globally warm climate.

These patterns should help us narrow our search for a cause (or causes) of the extinction since the effect on marine life, although massive, was largely restricted to particular kinds of organisms (although they were certainly not the only forms affected). Various plausible explanations for the extinction have been offered, however, so we need to assess each on its merits, bearing in mind the patterns noted above.

6.6.2 Plate tectonics revisited

We saw in Section 6.3.3 that plate tectonics can greatly affect climate in a number of ways. The assembly of the Pangean supercontinent and its position with respect to the South Pole, as well as volcanic activity and eustatic sea-level changes, all played some sort of role in the preceding icehouse world. Global climate became warmer by the mid-Permian and the ice sheets disappeared.

Since it seems that it was the organisms that lived on shallow continental shelves which were most affected at the end of the Permian, it is reasonable to look for a mechanism which directly explains the loss or restriction of this particular habitat. The most obvious is a global fall in sea-level, which would have resulted in the emergence of large areas of the shelf and a great reduction in the space available for these shallow marine creatures. Indeed, sea-levels declined through most of the Permian (punctuated by shorter-term rises and falls), culminating in a rapid **marine regression** (seaward retreat of shorelines) near the end (Figure 6.9), when sea-levels reached their lowest point of the whole Phanerozoic. They then rose rapidly during an early Triassic **marine transgression** (landward advance of shorelines), returning to their pre-latest Permian levels.

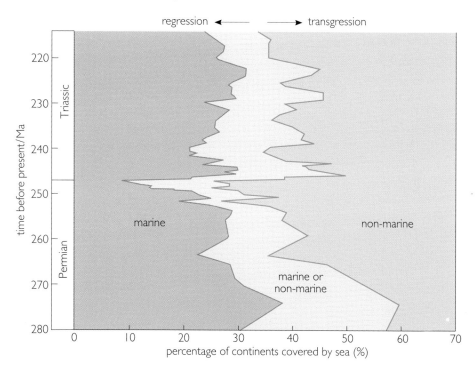

Figure 6.9
Changes in Permian and Triassic sea-level, expressed in terms of percentage of continental area covered by sea.

We have already looked at two long-term mechanisms of sea-level change in the Permo-Carboniferous – differences in mid-ocean ridge volume and changes in the extent of polar ice (Section 6.3.3). Since the formation of Pangea was accompanied by a reduction in the extent of mid-ocean ridge systems, ridge volumes would have been relatively low, leading to a fall in sea-level. On the other hand, the large southern ice-cap had already melted by the mid-Permian, and did not redevelop, so no part of the regression at that time can be attributed to that. Although melting of the polar ice sheet during the early Permian should have caused sea-levels to rise, any such tendency seems to have been more than offset by tectonically driven factors.

Although such tectonic processes can explain the gradual Permian regression, they cannot account for the much more rapid regression and subsequent transgression around the Permian–Triassic boundary itself. Changes in mid-oceanic ridge volumes tend to operate on a much longer time-scale. At least some of the change might be explained by looking at continental lava flows, particularly flood basalts, which you encountered in *The Dynamic Earth*. These are huge accumulations of lava produced by different eruptions over a short time. They are thought to be caused by hot spots (mantle plumes) and often signal the start of rifting and break-up of a continent.

Question 6.10
From what you have read in the Course so far, can you recall an example of a flood basalt that coincides in age with the Permian–Triassic boundary?

These basalts resulted from eruptions which appear to have occurred over only about a million years, totalling some 1.5 million cubic kilometres. Apart from adding more CO_2 to the atmosphere, the mantle upwelling and associated thermal doming of the crust may have caused a further short-term increase in land elevation, the net effect being rapid further regression. Once this activity ceased, subsidence would have lowered average elevation again, allowing the early Triassic transgression. It would also be tempting to relate the eruption of the Siberian flood basalts to the fall in the $^{87}Sr/^{86}Sr$ ratio near the Permian–Triassic boundary (Figure 6.3), but this is a little problematical.

▪ What would have been required for these continental basalts (with a low $^{87}Sr/^{86}Sr$ ratio) to have had such an effect on the strontium isotope ratio of marine limestones?

▪ Either a significant amount of basalt, additional to that remaining today, would have had to have been weathered and the products transported to the oceans, or there would have had to have been accompanying oceanic volcanism.

The lack of surviving ocean floor of this age makes testing the second possibility well-nigh impossible, but it seems unlikely, because the thermal doming of the ocean floor that would have accompanied any mantle plume development could be expected to have led to a global rise, rather than fall, in sea-level. We are thus left with the debatable possibility of extensive erosion of the continental flood basalts at the time.

These flood basalts *may* have contributed to atmospheric CO_2 buildup and global warming around the Permian–Triassic boundary (*The Dynamic Earth*), but there are other important factors to consider.

6.6.3 Another look at atmospheric CO_2, vegetation and climate

As you have seen in earlier chapters, changes in $\delta^{13}C_{carb}$ values through time reflect the relative burial rates of organic and carbonate carbon. As more ^{12}C is removed from the system (with burial of photosynthesized organic matter), so $\delta^{13}C_{carb}$ values in carbonates precipitated from the seawater increase. Conversely, a low $\delta^{13}C_{carb}$ value indicates low rates of organic burial relative to carbonate deposition. Figure 6.10 shows an accelerating decline in $\delta^{13}C_{carb}$ values during the latest Permian, into the earliest Triassic, with a subsequent increase, albeit interrupted by a second brief fall, later in the early Triassic. This could be interpreted as reflecting a decrease in the amount of organic burial relative to carbonate deposition at the start of the Triassic.

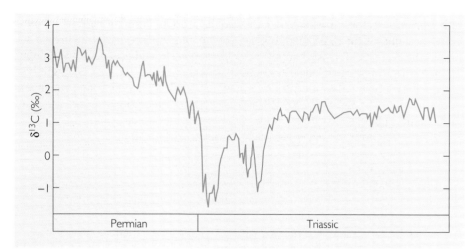

Figure 6.10
$\delta^{13}C_{carb}$ values around the Permian–Triassic boundary from limestones in the Carnic Alps of Austria. (The time-scale on this figure covers about 15 Ma.)

You should now refer back to the reactions discussed in Section 6.3.2, which show the ways in which CO_2 is removed from, or added to, the atmosphere over long time-scales. You saw that Permo-Carboniferous vegetation indirectly enhanced atmospheric CO_2 drawdown (and global cooling) through the increased weathering of silicates by carbonic acid. However, the low $\delta^{13}C_{carb}$ values seen at the start of the Triassic seem to imply some sustained carbonate deposition, with much decreased organic burial, at a time of greatly-reduced vegetational influence.

The continued sea-level fall, with erosion of previously lowland areas, may further help to explain the marked dip in $\delta^{13}C_{carb}$ values.

Question 6.11
How would the erosion of such areas in low latitudes have affected the flux of carbon between organic deposits and the atmosphere?

The CO_2 now released could have helped contribute to climatic warming as well.

The eventual increase in $\delta^{13}C_{carb}$ values later in the early Triassic could be taken to mean that the weathering and oxidation of buried plant remains decreased. Although the marine transgression at this time (Figure 6.9) may have had such an effect by once again submerging lowland areas, another contributory factor can also be considered. As you have just seen, low $\delta^{13}C_{carb}$ values in the carbonate record imply low rates of organic burial relative to carbonate deposition. So high values might be taken to mean the converse – high rates of organic burial relative to carbonate formation. There is another way of looking at this, however, which relates to the late Permian marine extinction. The main source of Phanerozoic carbonates is debris from the shells of marine organisms. These may well have been in anomalously short supply following the extinction event: the geological record at least suggests that the accumulation of such deposits took a while to recover to earlier levels. This deficit would have had the effect of reversing the tendency to low $\delta^{13}C_{carb}$ values produced by weathering and oxidation of land-derived organic matter. Any plankton blooms present at the time would have led to renewed organic burial, and hence further enhanced the increase. Note from Figure 6.10 that even when values did stabilize a little later in the Triassic, following a second excursion to low $\delta^{13}C_{carb}$ values, they did so at a lower mean level than that which had prevailed before the end-Permian collapse. The point to grasp here is that we are dealing with the net effects of two variables, either of which is capable of varying independently of the other: sequestering of organic, and of carbonate, carbon are each subject to a host of independent controls. Thus either may increase or decrease regardless of the other, giving a choice of possible explanations for any given change in $\delta^{13}C_{carb}$ values.

Could these biogeochemical changes around the Permian–Triassic boundary themselves help to explain the mass extinction then? Remember that the only significant source of atmospheric oxygen is through plant photosynthesis, and it is removed through respiration, rock weathering, and oxidation of reduced gases and ions, released by volcanoes and mid-ocean ridges. Therefore the amount of oxygen in the atmosphere at any one time is a function of photosynthesis minus the combined effects of respiration and geological oxidation (*Atmosphere, Earth and Life*).

Question 6.12

From what you have read in this chapter, how might the levels of atmospheric oxygen have been affected by conditions around the Permian–Triassic boundary?

The onset of such relatively oxygen-poor conditions, with associated oceanic anoxia, would have had a devastating effect on a wide range of marine creatures. Moreover, as equatorial conditions on land had already become more arid, there would have been a relatively reduced rate of organic matter production and burial, anyway (think of the loss of the Permo-Carboniferous swamps). This would have contributed further to increased atmospheric CO_2, reduced O_2 levels, and oceanic anoxia. Some evidence for this can be seen in Figure 6.11, which shows the changing abundance of coal-bearing sedimentary rocks (by area and by volume) since the Devonian (from the time when land plants started forming significant coal deposits).

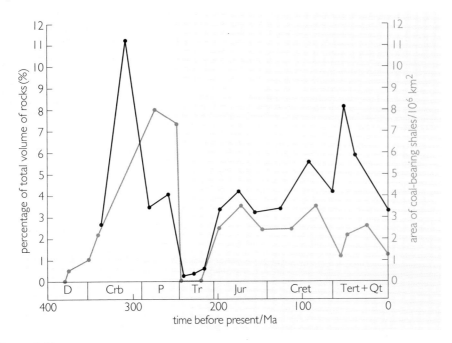

Figure 6.11

Changing abundance of coal-bearing sedimentary rocks through time. Note the high levels around the Permo-Carboniferous boundary and the sharp decline around the Permian–Triassic boundary.

The peak in both the volume of coal-bearing rocks and the area of deposition around the Permo-Carboniferous boundary supports the scenario set out earlier for the icehouse interval (Section 6.5). Note, though, the rapid drop in the amount of coal formation at the Permian–Triassic boundary, to the lowest levels recorded, and the subsequent earliest Triassic 'coal gap'. Further development of the monsoonal system led to increasingly arid conditions at low latitudes even in eastern Pangea. This caused previously unaffected vegetation (e.g. tree-sized lycophytes) to suffer. In high southern latitudes, the glossopterids also underwent a marked decline. There seems to have been a delay in response to these changes, such that it took time for the plants which colonized the new environments to diversify, stabilize soils, and contribute to peat/coal formation. Lowered sea-levels would have led to increased weathering and oxidation of buried organic matter (i.e. peat and coal). This may also explain why $^{87}Sr/^{86}Sr$ ratios rose again in the earliest Triassic (Figure 6.3), as a function of increased continental weathering. The oxidation of such large volumes of peat and coal would also have increased the level of atmospheric CO_2, and so contributed to warming. The lowering of CO_2 levels by continued collisional uplift (and subsequent weathering of exposed silicates) was apparently offset by the large increase in atmospheric CO_2 due to extensive oxidation of organic matter.

6.7 An extinction scenario

Given that the late Permian extinction occurred over some 3 (perhaps as many as 8) Ma, it cannot be described as an instantaneous or 'sudden' event. This means that we cannot invoke a single catastrophic mechanism to explain it. Indeed, Figure 6.12 (*overleaf*) illustrates the complexity of evidence and arguments used by various scientists to explain the extinction.

You have learned about a few of the factors that we consider to be more directly relevant to the story. Of those discussed, the closest we have to a 'sudden' event is the eruption of the Siberian flood basalts. But the importance of the basalts as a contributor to the extinction should not be overstated; they probably played a role (at least in contributing to atmospheric CO_2), but one which was secondary to another tectonically-induced mechanism – marine regression.

The marine regression had a number of consequences. A reduction in the continental shelf area would have caused major problems for the sessile bottom-dwelling marine creatures (the dominant members of the Paleozoic Fauna) living there, with a consequent reduction in their diversity. It would also have led to an increase in the weathering and oxidation of buried land-derived organic matter. The resulting release of CO_2 back to the atmosphere would have raised global temperatures. It would also have had the effect of lowering atmospheric O_2 levels, ultimately causing oceanic anoxia. Since this would have affected areas other than the continental shelves, it helps explain why marine organisms other than the dominant sessile ones were also affected, resulting in their partial extermination. This simplified scenario for the late Permian extinction is shown in Figure 6.13.

Of course, any proposed extinction scenario must be constrained by evidence, and the model suggested in Figure 6.13 is just one possible way of linking the late Permian extinctions to the timing and duration of possible contributory factors. Yet many of the proposed causes (some of which are shown in Figure 6.12) have gone unmentioned. It is only fairly recently that the majority of Earth scientists have 'taken a step back' to consider the broader implications of their highly specialized research findings. This has sometimes led to false trails, with global interpretations being made on the basis of only a few local observations. At present there is still no

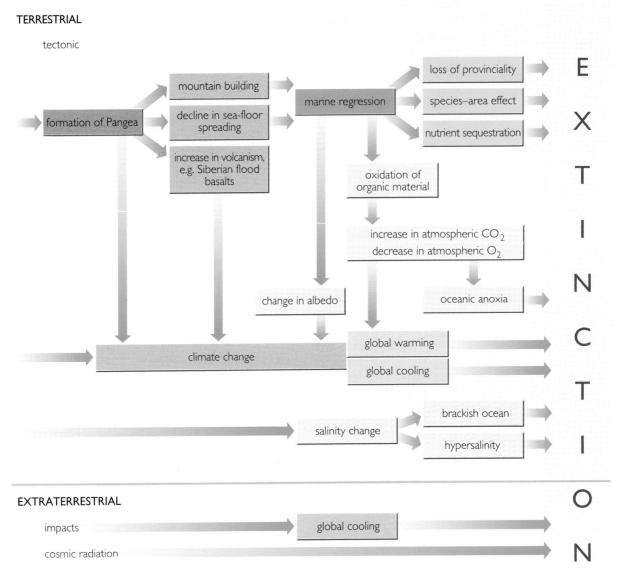

Figure 6.12
Some of the suggested explanations for the late Permian mass extinction, with the more indirect, though pervasive, factors shown to the left.

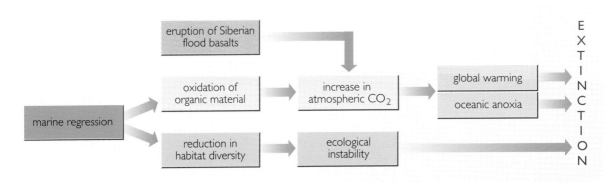

Figure 6.13
The simplified late Permian extinction scenario presented in this chapter.

clear consensus as to the relative importance (or even effect) of these various factors. But such a devastating mass extinction, spread over several million years, is likely to have been due to several interacting influences, rather than any single causal event.

6.8 Summary of Chapter 6

1   The Carboniferous icehouse world comprised one main continent, Pangea, part of which lay over the South Pole giving a 'cap world' configuration with extensive ice sheets in high southern latitudes. The main kinds of plants and animals were also different to today, although Carboniferous vegetational distributions were similar with plant productivity highest in low latitudes. There was extensive CO_2 sequestering and coal formation in these equatorial regions.

2   Other factors, such as extensive mountain building, relatively fewer mid-ocean ridges, marine regression, and even lowered levels of solar radiation, also contributed to the onset and persistence of globally cool conditions, either directly or by reducing atmospheric CO_2 levels.

3   The onset of globally warm conditions in the Permian led to a loss of the ice sheets. Patterns of vegetation also changed, with the loss of the equatorial swamp forests of Europe and North America and the development of forests in high southern latitudes. The movement of Pangea away from the South Pole and a developing monsoonal system contributed to the overall warming. Increasingly arid conditions at low latitudes led to (a) the exposure and oxidation of previously-sequestered carbon, releasing CO_2 back to the atmosphere, as well as (b) less vegetation and consequently reduced rates of atmospheric CO_2 drawdown.

4   The late Permian extinction had a devastating effect, not least on marine life. It was not a 'sudden' event, since it occurred over some 3 to 8 million years although recent work suggests two discrete episodes of extinction. Sessile shallow marine organisms suffered most, due to a reduction in shelf area caused by marine regression. This also affected other forms of life, since regression led to increased coal weathering, release of CO_2, ocean anoxia and further global warming.

5   It is the relative importance of different factors, as well as the interplay and feedbacks between different processes which must be considered. For example, marine regressions occurred both in the late Carboniferous and latest Permian, but with two different effects. The first enabled vegetation to expand and colonize the newly-exposed land, enabling further CO_2 sequestering and cooling, whereas the second led to weathering of previously-buried coal, release of CO_2 and enhanced global warming. This can be explained by changes in the relative importance and role of other factors, such as the development of a monsoonal system and changes in the kinds of plants able to colonize these environments.

6   Interpretations are further complicated by uncertainties concerning the timing of events and processes. A complex interplay of interactions and feedbacks is suspected to have been responsible for both the icehouse–greenhouse transition and the late Permian mass extinction.

Chapter 7
A greenhouse case study

7.1 Introduction

As the break-up of Pangea proceeded through the Mesozoic, the climatic warmth established in late Permian times (Chapter 6) persisted.

If you look back to Figure 5.1 you can see that the warmest interval in the last 443 Ma was in the Cretaceous. We will therefore use this interval as a case study of what an extreme greenhouse world can be like.

As in the previous chapter, we shall start by looking at the geography of the times (in Section 7.2) then move on to the land plants, with special emphasis on those that lived at high latitudes (Sections 7.3–7.5). Next, following an interlude on climate modelling (Section 7.6), we turn to life in the sea (Section 7.7) before drawing together the threads to provide a model for the greenhouse Cretaceous world (Section 7.8). Finally, in Section 7.9, some broad conclusions from both the case studies in Chapters 6 and 7 are noted.

7.2 Cretaceous geography

Figure 7.1 shows the reconstructed arrangement of continents for mid-Cretaceous time and the distribution of some climatically-sensitive deposits (as discussed in Section 6.2.2). The widespread occurrence of coals at high paleolatitudes (>60°) demonstrates that temperatures were high enough there for plant life to flourish and that large volumes of glacial ice were absent. At low paleolatitudes (<30°) the extent of evaporites (especially in the Northern Hemisphere) shows that large areas were predominantly arid, with evaporation exceeding precipitation. Contrast this with the extensive tropical rainforests of today in the Amazon Basin, Africa and South-East Asia. Although some coals (a sign of active plant growth and therefore water) formed at mid- to low paleolatitudes (<60°) in the Cretaceous, they did so on a limited scale and only close to ocean margins.

Another noticeable feature of the map is that the individual landmasses are smaller in extent than those of today: there are no corresponding areas of land as large as the modern Eurasia or North America, for example. The reason for this is two-fold. First, if you refer back to the map showing latest Permian geography (Figure 6.1b) you can see that the Cretaceous world was a product of the rifting apart of the large supercontinent Pangea. Second, Cretaceous sea-levels were, at times, a few hundred metres higher than at present, so that large areas of the continents were covered by shallow seas.

Question 7.1
From what you now already know about the differences between the Cretaceous Earth and the Permo-Carboniferous icehouse Earth, can you think of two reasons why sea-levels should have been relatively higher in the Cretaceous?

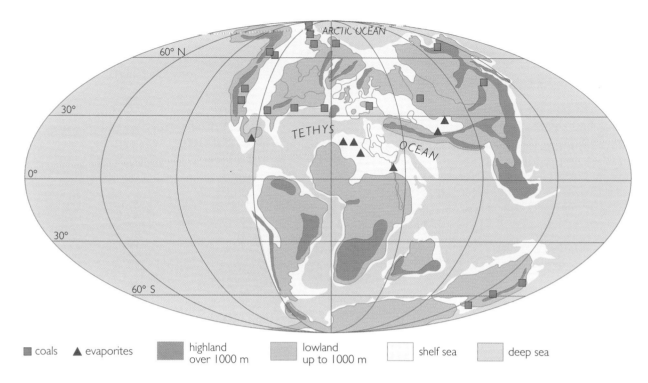

Figure 7.1
Mid-Cretaceous paleogeography (around 95 Ma ago) and climatically-sensitive deposits.

One of the shallow continental interior seas, for example, was the Western Interior Seaway in North America, which at various times during the Cretaceous connected the Tethys Ocean with the Arctic Ocean. These shallow seas were important to the world climate in that:

◆ they were a source of moisture in what would otherwise have been dry land areas;

◆ they warmed and cooled more slowly than the surrounding land (because of the high heat capacity of water) and therefore affected heat distribution; and

◆ they acted as conduits for heat as water currents flowed from one ocean to another.

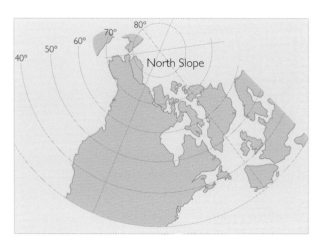

Figure 7.2
Mid-Cretaceous position of Alaska.

7.3 Polar climate

7.3.1 The case of Alaska

Climate change is mostly strongly expressed at the poles and for this reason we will begin to look at the greenhouse world of the Cretaceous by studying the evidence for its polar climates, and in particular we shall look at the high Arctic.

One of the best known Cretaceous sedimentary sequences in the Arctic is that of northern Alaska. Today, the northernmost point in Alaska is Point Barrow at 71.23° N. In Figure 7.2 you can see the position of northern Alaska (the North Slope) on a reconstruction of a part of the mid-Cretaceous Northern Hemisphere.

■   What was the paleolatitude of northern Alaska in the mid-Cretaceous, and is that northward or southward of its present latitude?

■   It was between 75 and 85° N in mid-Cretaceous times, which is further north than its present position.

Figure 7.3 shows the distribution of light and darkness at high latitudes today, with the obliquity of the Earth at about 23°. The obliquity of the Earth is measured as the departure of the axis of rotation from the vertical, where the 'vertical' is defined as being perpendicular to the Earth's orbital plane around the Sun. You can see that, at the latitude of Point Barrow today, continuous winter darkness (i.e. darkness longer than 24 hours) does not occur, but there are about 2 months of twilight. For a period of almost 4 months, day-length changes until during the summer there is a period of about 2.5 months when there is continual daylight.

Figure 7.3
The distribution of light and darkness at high latitudes today. The colour bands represent the different durations of daylight (in hours) over any given 24 hour period.

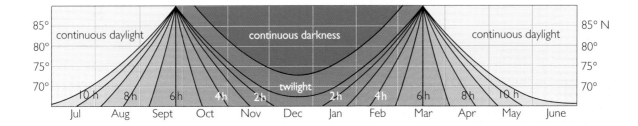

■   What is the duration of continuous winter darkness at 80° N?

■   About 3 months.

Northern Alaska, and the Cretaceous rocks found there in particular, is extremely rich in coal. In fact, current estimates suggest that there are 2.75×10^{12} tonnes of coal lying under northern Alaska – this represents about one-third of the total US coal reserves of all ages combined (including those of the Carboniferous). By any measure, then, this near-polar environment was an extremely effective long-term carbon sequestering and storage system during the Cretaceous.

We know that the coal-forming vegetation was in the form of forests rather than tundra, because tree stumps are preserved in life-position rooted in fossil soils. Their existence poses two questions. First, why was this forest ecosystem so effective at capturing carbon and, second, how was the carbon so effectively buried?

It is easier to answer the second question first. Rocks of mid- to late Cretaceous age were laid down in northern Alaska as a series of deltaic sediments that built out northwards into a trough to the north of the newly uplifted Brooks Range mountains. Though the details are unresolved, the mountain range appears to have been generated by local plate collision. Uplift was most rapid during the last part of the early Cretaceous, but continued through into the late Cretaceous. Erosion of these mountains during the mid- to late Cretaceous supplied the deltaic sediments.

Plants colonized the delta floodplains skirting the mountains and, as the sediments containing the organic matter subsided (under gravity), more sediments were deposited on them, and more forests grew. Such a process, typical of many long-lived deltas, provided an ideal setting not only for the preservation of fossils but also for burying peat.

The first question on carbon capture is more complicated to answer and it is to that issue that we now turn.

7.3.2 The Arctic forests

There are peats forming in northern Alaska today, but there are no forests. Since, in the Cretaceous, northern Alaska was even more poleward of its present position, we are forced to conclude that climatic conditions there were not the same in the Cretaceous as they are now. Conditions must have been more benign in the Cretaceous to have allowed the growth of forests at these latitudes. Or were they?

▪ What alternative explanation might there be for Cretaceous polar forests if climate conditions then had been the same as now?

▪ The tolerances of plants might have changed. For example, the plants might have had a different biochemistry that allowed enzymes, etc. to function more efficiently at low temperatures.

This is unlikely, however, because one might expect plants with such specialized adaptations to have survived subsequently, particularly during the Quaternary glacial events, and to still be in evidence today. The fact that such plants are not still with us, particularly in today's cool climate, suggests they never existed.

So if conditions (rather than the plants' fundamental requirements for growth) were different, how can we deduce what those conditions were like? To do that we must examine the rocks and try to decode the information they contain. We must examine not only the fossils but the sedimentary context in which they are found.

Plant fossils have been known from the Arctic for a very long time. Some finds of apparently warmth-loving plants at high latitudes can be eliminated from further consideration because they are found on tectonic plates that were at lower latitudes when the plants were alive, but which subsequently drifted to their near-polar settings of today. We are still left with the problem of northern Alaska, however, the paleolatitude of which for the Cretaceous is well constrained (to within 5°). A wealth of fossils can be found there, but the fact that they represent trees is incompatible with present conditions at the latitudes where they grew.

7.3.3 The cycad conundrum

Among the more problematical finds are those that are thought to be fossil **cycads**. These plants, that naturally occur today only at low latitudes where frosts are absent or infrequent, are represented most commonly in the Cretaceous fossil record of Alaska by the leaf form given the generic name *Nilssonia* (Figure 7.4).

Modern cycads typically have the appearance of a squat palm tree: they have a trunk composed mostly of tissue rather like that in an apple or potato, and very little wood. At the top of the trunk is a crown of leathery evergreen fronds (Figure 7.5). The sensitivity of the modern plant to frost, and its evergreenness, have posed severe problems for interpreting Arctic paleoclimates. For a start, if the biology of Cretaceous cycads had been the same as that of the modern ones, the polar regions would have to have been much warmer than now. It is difficult to see how such conditions could have been sustained with long periods of winter darkness.

Even more problematical, however, is the fact that modern cycads retain their leaves all the year round: they are evergreen. In addition to photosynthesizing, plants must respire, to release the energy stored in their food reserves. The rate at which metabolic processes, including respiration, proceed increases with temperature. The optimum temperature range varies with the type of plant, but in general most species function best between 5 °C and about 40 °C. Below 5 °C,

├────┤ 1 cm

Figure 7.4
Nilssonia frond.

Figure 7.5
Modern cycad, *Cycas*. The plant is about 2.5 m in height.

metabolic rates of all living things are very slow (that is why domestic refrigerators work at about 4 °C – bacterial action is curtailed at this temperature), while above 40 °C some processes are inhibited or prevented as enzymes start to become damaged. Thus the balance of photosynthesis and respiration is both light-, and temperature-dependent. In climates where temperatures never, or only very rarely, drop to near freezing, respiration rates are relatively high. In cold climates, especially in winter, respiration rates are very low, so respiration in the leaves does not use up much food. Plants in cold dark situations, such as are experienced today during winters at high latitudes, can survive, even though the darkness prevents the production of food by photosynthesis. This is because they do not exhaust their food store since the rate of respiration is also low.

Question 7.2

If the conditions were warmer at the time, why should the possibility of evergreenness at high latitudes in the Cretaceous be problematical?

Here we have assumed that in the late Cretaceous the Earth's obliquity was the same as at present – a point to which we will return below. Given that the paleolatitude of the centre of the North Slope was 80° N, it follows from Figure 7.3 that the *Nilssonia* plants would have experienced continuous winter darkness for some 3 months, and that in the spring and autumn there would have been a period of 2 weeks of twilight. If the *Nilssonia* plants were indeed evergreen, maintaining the leaves through a long, dark but relatively warm winter, would have caused an intolerably high respiratory drain on the plant, particularly for young plants and seedlings.

It is not surprising, then, that some scientists have suggested that if the Arctic cycads were evergreen, then they must have received winter light so that they could manufacture food to replace some, or all, of that consumed by leaf respiration. To achieve this more even distribution of light all the year round at the poles, the obliquity of the Earth would have had to have been considerably less than the present 23.4°: to have allowed evergreens to grow at such high latitudes, they argued, the Cretaceous obliquity might have had to have been as low as 5°. This reduction is of course outside the range produced by the Earth's regular orbital variations (*The Dynamic Earth*) and would have required a special mechanism to have effected the change. Yet, although other planets in the Solar System, Mars and Uranus, may have experienced even larger obliquity changes in the past, possibly caused by asteroid impacts, the stability of the coupled Earth–Moon system is regarded as making such an effect exceedingly unlikely.

Are there any additional factors that we need to take into account before we become concerned about the apparent anomaly of evergreen plants growing in northern Alaska in the Cretaceous?

One thing we have overlooked so far is the question of whether we can be sure that the paleomagnetic pole used to determine the paleolatitudinal positions of the continental plates was the same, or nearly the same, as the Earth's rotational pole. It is the position and inclination of the rotational pole or axis, of course, that determines the light regime. To examine this question we can look at the symmetry of distribution of climatically-sensitive deposits. These sediments ought to be symmetrical about the rotational pole because the rotation of the Earth controls atmospheric dynamics and therefore climate. If the position of the sediment-determined rotational pole agrees with that determined paleomagnetically then we can assume that the rotational and magnetic poles were essentially the same.

In fact, in the latest Cretaceous, when northern Alaska was at paleomagnetic 85° N, the rotational pole appears to have been within 4° of the paleomagnetic pole, i.e. less than the error within which we can position the pole using paleomagnetism. We may therefore conclude that the rotational and magnetic poles were effectively the same at that time, and that the paleomagnetically determined paleolatitude may be assumed to have been more or less the same as the true paleolatitude in relation to the rotational pole. Therefore, we cannot escape the conclusion that these cycads did indeed experience prolonged winter darkness.

Has the biology of the plants been correctly interpreted? Modern cycads exhibit what is known as a relict distribution. That is to say that they are restricted to relatively few sites that are geographically isolated, and that individual genera occur at several of these separated sites. The more widespread distribution and greater taxonomic diversity of cycads in the Mesozoic implies that their present distribution must represent merely a relict of the former glory of the group as a whole. This poses several questions:

1   Were the environmental conditions that suit cycads more widespread in the Mesozoic (i.e. was global climate more uniformly warm)?
2   Or are the present, relictual, forms adapted to a narrower range of conditions than that tolerated by Cretaceous forms?
3   Or has the environmental tolerance of cycads become more restricted?

To answer these questions we need to go back to the fossils themselves and examine them without being prejudiced by the biology of the modern relicts. In 1975 two Japanese paleobotanists, Tatsuaki Kimura and Shinji Sekido, described a Cretaceous cycad quite unlike any that are alive today. They called this plant

Nilssoniocladus because it bore *Nilssonia* leaves branching (Greek, *klados*, a shoot) from side shoots (Figure 7.6).

Unlike typical modern cycads, which have a squat palm-like appearance with a thick trunk, *Nilssoniocladus* had a thin vine-like stem. Arranged along this stem were side shoots which bore numerous scars where leaves had been attached. This observation suggests that *Nilssoniocladus* was a deciduous cycad – one that shed its leaves seasonally throughout its life. Moreover, the plant apparently not only shed its leaves seasonally but did so in a synchronous fashion (i.e. leaves were shed at the same time, leaving the plant devoid of all leaves for significant periods in its life). Evidence for this comes from the facts that *Nilssonia* leaves are often found as leaf mats and that individual leaves appear not to have broken off from the main plant but have distinct bases specialized for leaf separation. Moreover they are usually found intact and therefore must have been shed intact, without damage or rotting. Leaves of different sizes are found on single bedding planes indicating that they all fell off the plant at the same time and not just when they were full-sized. Because of this powerful evidence that *Nilssoniocladus* was deciduous, we can now infer that it was the biology of the Cretaceous cycads that was different. The modern cycads thus represent a relictual subset of a group that previously had had a wider range of environmental tolerances.

Figure 7.6
A reconstruction of part of the *Nilssoniocladus* plant that grew throughout the mid- and late Cretaceous of the Arctic. The fronds are about 20 cm long.

Notice here that we are using a combination of sedimentological evidence, observations on leaf characteristics (size, shape and whether they display signs of decay) and shoot architecture. The study of sedimentological and biological evidence relating to explanations of fossilization phenomena is known as **taphonomy** (from the Greek *taphos*, a tomb) and is of critical importance in paleoenvironmental research.

7.3.4 A new perspective

What are the implications of this new understanding of Mesozoic cycads? Well, first, that it is dangerous to extrapolate uncritically the biology and climatic tolerances of modern plants back in time, particularly beyond the Quaternary. Second, if the cycads were deciduous, then the 'problem' of winter darkness ceases to be a problem, at least for them. But was deciduousness a general characteristic of the Cretaceous vegetation of northern Alaska?

In fact there is similar evidence for deciduousness in many other plant groups. Typically the angiosperms, for example, first appear in the middle part of the Cretaceous succession, with large leaves that look very much like those of modern plane trees (Figure 7.7). These are informally termed platanoid leaves. Platanoid leaves are typical of deciduous angiosperms: they tend to be cheaply constructed in that they are thin-textured. This is in contrast to longer-lived, thicker, leaves produced by evergreens, such as rhododendron or holly. Moreover, platanoid leaves

Figure 7.7
A Platanoid leaf in sandstone
(× 0.5).

have relatively thin cuticles, because they do not have to limit water loss during the winter, when the root zone might be frozen. They also have an expanded base to their stalk – a feature of seasonally shed leaves.

When we say the platanoid leaves are large, we should say that the size range is large: there are some small ones. However, many are more than 10 cm in breadth. In spite of their large size, and the coarseness of the sediment, these leaves are invariably complete within the rock. As in the interpretation of leaf mats, this strongly suggests that they were shed intact and were all buried together before any of them had had time to become mechanically weak through decay.

So far we have considered representatives of angiosperms and cycads, but these represent only a small fraction of the polar land vegetation. The most ubiquitous plants appear to have been deciduous conifers similar to modern redwoods, together with ginkgos, or 'maidenhair' trees.

The options of 'deciduousness' or 'evergreenness' are restricted to long-lived woody taxa; herbaceous plants do not normally shed their leaves. What did they do during the winter darkness?

Much depends on whether the plants were annuals or perennials. Annuals grow from seed each year, reproduce and then die in the autumn. Overwintering is in the form of seeds. Perennial herbaceous plants adopt a different strategy; they die back to basal food storage organs such as rhizomes (a modern example is the iris), corms (e.g. crocus) or bulbs (e.g. tulips or onions). In the Cretaceous we have no evidence for corms or bulbs, but many ground-cover plants had rhizomes. We can find remnants of these rhizomes in fossil soils. Many are characteristic of modern *Equisetum* (horsetails or scouring rushes), and are assigned to the fossil form *Equisetites*, while others represent ferns. In some instances it is evident that ferns formed the ground cover because they are found preserved in place, buried beneath volcanic ash-falls overlying fossil soils. In the winter, then, the above-ground parts of *Equisetites* and ferns, the two main ground-cover plants (fossil evidence suggests that grasslands had not yet evolved) would have died back to their rhizomes and become dormant.

▨   What do you think might have become of the organic carbon making up those seasonal above-ground parts?

▨   As with many grasses, some would have been eaten by animals, but most would have eventually rotted or become buried by sediments.

Fossil evidence indeed shows that the near-polar forests also supported a diverse fauna, including large herds of plant-eating dinosaurs such as the duck-billed *Edmontosaurus* or the horned ceratopsians, which were preyed upon by meat-eating dinosaurs such as *Tyrannosaurus*.

Wildfires were also common in the polar forests (there is abundant fossil charcoal) and must have converted some plant carbon back into CO_2, while yet more would have been oxidized through bacterial action or converted into methane under reducing conditions in boggy areas. Other remnants of the vegetation would have been transported as partly-rotted microscopic particles into sediments where they have survived as particulate material or as organic molecules.

From the kinds of evidence discussed above, then, we can build up a picture of the northern Alaskan Cretaceous forests as consisting of plants that were deciduous or that died back to underground organs during the winter months. There is no need to postulate year-round light, or day/night cycles.

There is further, direct, evidence that the polar light regime was like that of today. In addition to fossil leaves, the Cretaceous sediments of northern Alaska yield abundant fossil wood with well preserved internal structure (see Box 6.1). In some sites in northern Alaska, tree stumps are even found rooted in fossil soils (Figure 7.8), so we know that trees did indeed grow in those near-polar conditions. When thin sections of the fossil logs are studied under the microscope, fine preservation of structure, even at the sub-cellular level, may often be seen (Figure 7.9). However, interpretation of these preserved structures is not straightforward. A given pattern of ring features may be the result of genetic factors rather than environmental constraints. Yet if a number of different taxa display the same pattern, it is probable that the pattern is environmentally produced, either through direct influence on growth, or through selection for certain genetically produced characteristics that confer specific advantages in the prevailing environmental regime.

Figure 7.8
Photograph of a mid-Cretaceous fossil tree stump in life position. This tree fossil was found in Alaska at a present latitude of 69° N but when alive it would have been growing at about 80° N as part of a polar forest.

The pattern observed in many of the northern Alaskan woods of mid-Cretaceous age is like that shown in Figure 7.9. The earlywood produced in the spring is clearly visible as xylem cells with large internal cavities, through which water flowed. The cell walls of the earlywood cells are thin. The latewood cells have thicker walls and smaller cavities, forming the dark bands in the wood.

Question 7.3
What does the large ratio of earlywood to latewood cells in Figure 7.9 tell us about the polar light regime under which the tree grew?

The uniformity of the mid-Cretaceous earlywood shows that growth conditions varied little throughout the growing season. This principally means that drought (or waterlogging) and chill rarely occurred. It also means that other events that could diminish growth, such as severe insect attack, were also of no great consequence. Towards the end of the Cretaceous, however, environmental conditions changed markedly. Figure 7.10 illustrates a cross-section of wood from some latest Cretaceous non-marine sediments, deposited at 85° N paleolatitude.

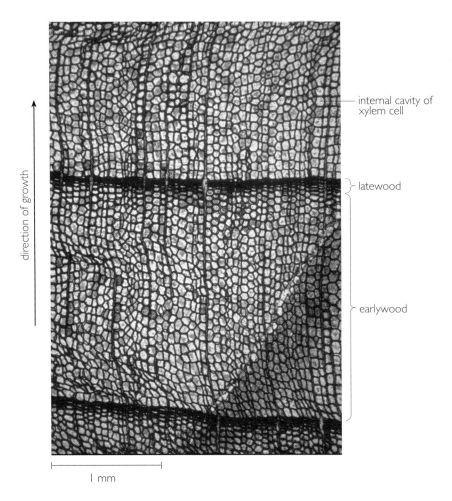

internal cavity of
xylem cell

latewood

earlywood

1 mm

Figure 7.9
Transverse section across the
wood of *Xenoxylon
latiporosum*; a coniferous tree
from the mid-Cretaceous of
northern Alaska.

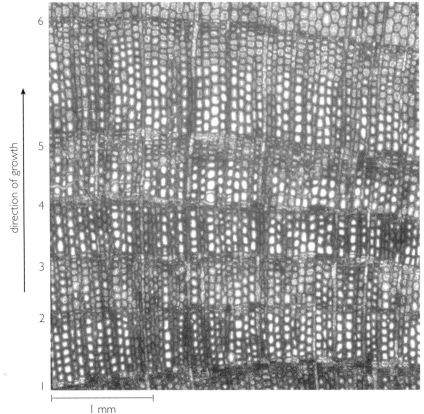

6

5

4

3

2

1

direction of growth

1 mm

Figure 7.10
Transverse section of wood
of latest Cretaceous age,
from northern Alaska.

Question 7.4
In what ways do the rings of the specimen in Figure 7.10 differ from those seen in Figure 7.9?

The upper two rings (numbers 5 and 6) show the normal characteristics of an annual ring boundary, where progressive slowing of growth is followed by dormancy and a sudden burst of growth the following spring. The lower rings (2 and 4) reflect interruptions in growth that occurred during the summer growth period. These interruptions can be caused by any factor that reduces the tree's photosynthetic activity. For example, they might be caused by sudden cold snaps that depress the rate of photosynthesis, drought which causes the stomata to close as a water conservation measure, thereby stopping the supply of CO_2 to the photosynthetic process, or by trauma such as massive insect damage.

Such rings are called **false rings** and are a measure of environmental variations that exceed the tolerance of that particular plant.

Many of the latest Cretaceous woods are characterized by frequent false rings. Often, false rings and true (annual) rings are difficult to differentiate and this in turn makes measuring ring thicknesses difficult. In general, however, these later trees have narrower rings, but because the amount of latewood is about the same as that in the mid-Cretaceous specimens the ratio of earlywood to latewood diminishes. This implies that growth conditions throughout the latest Cretaceous summer were less benign than those in the mid-Cretaceous.

Examination of the sediments reveals that drought indicators such as preserved mud cracks are lacking, although the frequency of charcoal in the latest Cretaceous sediments is higher than in the earlier beds, which implies that vegetation was drier and so more predisposed to combustion. Therefore, in the absence of direct evidence for insect damage (e.g. chewed leaves, or borings in the wood), the most likely explanation for the variation in summer growth was temperature fluctuation: there were periodic cold snaps. This might imply that the average summer temperatures were lower in the latest Cretaceous, such that temporary falls in temperature from the lower average value were sufficiently cold to limit photosynthetic activity.

Question 7.5
What two explanations could you offer to account for this cooling?

7.3.5 Taking stock

So, using qualitative sedimentological and paleobotanical evidence only we have established the following points:

1   The obliquity of the Earth's rotational axis was essentially the same in the late Cretaceous as it is now.
2   The rotational pole was essentially the same as the magnetic pole.
3   The polar light regime was similar to that of today.
4   The Arctic climate supported forests up to at least 85° N.
5   The overall polar climate must have been warmer than now.
6   There was sufficient rainfall to support abundant plant growth.
7   The vegetation must have dried sufficiently to burn from time to time.
8   The Arctic forests were deciduous.
9   Some plants had substantially different climatic adaptations when compared with their modern relatives.

By now we have established quite a detailed picture of the Cretaceous Arctic. We don't have time to investigate the Antarctic in similar detail, but there is substantial evidence to show that in many respects the high southern latitude vegetation and climate mirrored that of the Arctic.

7.4 Low-latitude vegetation and climate

7.4.1 Water supply

Now we turn our attention to the other end of the Equator-to-pole temperature gradient that powers the climate system: the low latitudes. If we return to Figure 7.1 for a moment, you will recall that one feature of the Cretaceous low latitudes was the extensive area of arid climates characterized by evaporites. To form, evaporites need not only a high evaporation rate but they also have to have a supply of water carrying dissolved minerals, so that continued salt precipitation takes place. The presence of evaporites over large continental areas therefore tells us that those areas were not constantly dry but must have had some water supply on a frequent basis. If this were so, then we might also expect to see some evidence for vegetation. Fortunately this is exactly what we do see and, as before, the plants tell us a great deal about the environment.

In Cretaceous sediments deposited between paleolatitudes 40° N and 40° S it is common to find large quantities of the fossil pollen type called *Classopollis*. This is a very easily recognized pollen form that is known to have been produced by a family of extinct conifers known as the Cheirolepidiaceae. The pollen is found in the cones of these plants, for which fossils of both foliage and wood have also been identified. We can find examples of cheirolepidiaceous foliage and wood in many early Cretaceous sediments in southern England (then around 36° N), deposited on the borders of a seaway connecting with the Tethys Ocean. On the Isle of Portland, for example, tree stumps, logs, foliage and pollen are found associated with sediments of earliest Cretaceous age containing **halite** (salt) **pseudomorphs** – casts of sediment filling voids from which the original halite crystals had been dissolved. This and other evidence suggests the plants grew under considerable evaporative stress where rainfall was, at best, highly seasonal. There is also a strong seasonal signal in the growth rings of the wood.

Different species of the Cheirolepidiaceae are found in the early Cretaceous sediments of lower latitudes. Again, wood, foliage, cones and pollen are all present, and the foliage form considered here, called *Frenelopsis*, shows some extremely interesting adaptations to aridity.

Figure 7.11a shows a foliage branch of a typical member of the Cheirolepidiaceae, many species of which have been reconstructed as a tree with a single main trunk. The foliage shoots are segmented with each successive segment inserted in the previous one rather like a stack of paper cups. The leaves are not immediately obvious but are small triangular shaped appendages situated at the rim of each shoot segment. In the species shown here, there are three leaves per segment but some species have only one.

Each segment is covered in rows of tiny dots. These are in fact the stomata, large numbers of them on each segment. A single stoma of a related genus *Pseudofrenelopsis* is illustrated in Figure 7.11b and it consists of a ring of finger-like projections that point towards each other over a deep pit in the thick cuticle. At the bottom of this pit is the stomatal aperture, bounded by a pair of guard cells (Figure 7.11b). The thick cuticle immediately provides a clue as to the functional significance of this elaborate stomatal structure: it reduced water loss.

(a)

Figure 7.11

(a) *Frenelopsis ramosissima* shoot showing the typical segmented structure seen in many members of the extinct Cheirolepidiaceae (× 10). (b) Scanning electron micrograph of a single stoma of *Pseudofrenelopsis* (× 2000). (c) Vertical section through the cuticle of *Pseudofrenelopsis* showing the deep stomatal pits (× 250). (d) Leaf of *Pseudofrenelopsis* showing marginal hairs behind which is the groove at the junction with the next shoot segment (× 60).

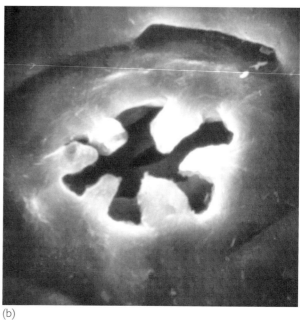

(b)

(c)

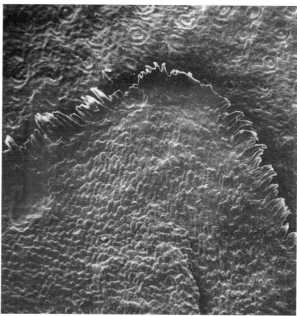

(d)

A thick cuticle is usually a good indicator that the plant producing it was exposed to considerable water stress, but this stomatal architecture represents a further strategy for resisting desiccation. The deep pits (Figure 7.11c) would have contained relatively static air which would have become water-saturated, so protecting the guard cells from drying. In effect, this adaptation is a 'captured' boundary layer (Section 4.6.2). The guard cells could only have had a thin cuticle if they were to retain the ability to change shape to control the stomatal apertures and they were therefore particularly susceptible to desiccation. Furthermore it seems likely that, at times of extreme drought, loss of water from the cells supporting the finger-like projections would have caused them to close up, so sealing the tops of the stomatal pits and protecting the guard cells from drying.

Another adaptation to drought conditions is seen in the architecture of the junctions between foliage shoot segments. This is shown in Figure 7.11d. At the top edge of the lower segment, the cuticle was fringed with a line of hairs that lined the edge of the groove between the segments. Within the groove the cuticle was very thin compared to that directly exposed to the atmosphere. We cannot be sure of the significance of this structure, but it is likely that the hairs would have been the sites for the nucleation of water droplets when dew formed or when it was foggy. The same phenomenon is seen when you hang a woollen sweater out overnight in damp air – the hairs of the sweater become covered in droplets of dew. If this did happen with *Pseudofrenelopsis*, then the droplets would have grown until they touched the cuticle of the adjacent segment, and the water would have been drawn into the thinly cutinized groove between the segments and absorbed by the plant. This plant may well have had the ability literally to extract water from the air, as mist and fog rolled in from the nearby sea.

Question 7.6
Now identify five adaptations to survival in an arid environment that are displayed in the fossilized remains of *Pseudofrenelopsis*.

The Cheirolepidiaceae were not the only plants adapted to this harsh environment. There were also some ferns with very thick cuticles and specialized stomata to combat water loss. Additionally, parts of the fern fronds are often found only as charred remains, showing that the vegetation was frequently burnt and must therefore have been dry for this to have happened.

7.4.2 Qualitative and quantitative inferences

Clearly the low-latitude Tethyan margin plants were architecturally very different from those growing in the polar regions, and these differences tell us a great deal about the climates of those regions, particularly temperature, evaporative stress, and frequency of rainfall. We can also tell that the main sites of land-based plant productivity were at high, not low, latitudes and that the equivalent of today's tropical rainforests did not exist to any great extent in the Cretaceous. These observations are of critical importance to understanding what might happen to climate in general, and agriculture in particular, should the Earth return to its normal warm (greenhouse) condition. When we combine these observations with those of sediments we learn even more, but all this information is qualitative. We cannot say what the exact temperature was, or how much rainfall there was. Methods of obtaining quantitative paleoclimate data are required.

7.5 Quantifying climate reconstructions

7.5.1 Plants as thermometers

So far we have reviewed the evidence that suggests that the late Cretaceous polar light regime was more or less as it is now. However, apart from deducing that the overall temperature regime was warmer than now, but need not have been very warm, and that frosts were likely, we have not really got to grips with quantifying the temperature regime. Here we will be considering some of the ways in which atmospheric Cretaceous polar temperatures might be determined. In fact, the same techniques may be applied anywhere that suitable plant fossils are preserved.

In Section 7.3.2 we deduced that it would be unreasonable to invoke special photosynthetic chemistries to explain high-latitude floras, so we will assume that the range of temperature within which growth could occur was much the same as it is today, somewhere between 5 °C and 40 °C. This is a rather large temperature range and little help in trying to understand the dynamics of polar (and global) climate. Paleobotanists have long recognized that land plants are excellent climate indicators and, broadly speaking, there are two approaches commonly adopted.

7.5.2 The problem with relatives

The first, and oldest, technique is based on the climatic tolerances of *nearest living relatives* (NLR), and is known as the **NLR technique**. This assumes that ancient plants and plant communities lived under similar conditions to those of their nearest living relatives. The success of the technique, which is widely used in Quaternary studies, depends on the correct identification of the fossil and, of course, an absence of evolutionary change. Where the fossils are in the form of reproductive structures, such as pollen or seeds, NLR techniques are particularly useful because the taxonomy of living plants is based mostly on their reproductive characteristics; correct identification is less difficult than it is with vegetative organs such as leaves. However, for pre-Quaternary fossils that may represent extinct taxa with very different environmental tolerances to their living relatives, and for vegetation for which there are no living counterparts such as the greenhouse polar forests, we have to adopt a different approach.

▪ Name an example of such a problem that you have already encountered.

▪ In Section 7.3.3, you read about the Cretaceous high-latitude cycad, *Nilssoniocladus*, which turned out to have been deciduous, in contrast to its evergreen living relatives.

7.5.3 Following form

The second method for using plants as climate indicators is based on the architecture, or **physiognomy**, of the plant or community, and is applied when the fossils are in the form of vegetative organs, particularly leaves. Because a plant cannot move around once it has taken root, it has to be well adapted to its local environment or it will die, either because its environmental tolerance is exceeded, or because it is out-competed by better adapted plants. Either way, there is a selective premium on being as efficient as possible in the local circumstances with respect to water conservation, gas exchange and light interception. In the course of evolution, most plants have become honed for successful exploitation of particular environmental niches and many display specially adapted physiognomies. One extreme example is seen in desert plants which have a low surface area to volume ratio such as to conserve water, low leaf-area indices (Section 4.5), and thick cuticles.

- ▦   Can you think of some examples?

- ▦   Cacti, and their Old-World counterparts, euphorbias, illustrate these traits well.

Rainforests, by contrast, are characterized by plants with large leaf-area indices, forming a vertical succession of layers within the forest.

Within particular habitats, vegetative organs can also display a certain degree of variation in form allowing them to be fine-tuned to local conditions. An example of this is seen where leaves at the top of a tree crown, exposed to high light levels but potentially desiccating winds, are small in area but thick, with robust cuticles. Those leaves on the same tree that are borne beneath the canopy in shaded, humid, conditions are large and thin, with only limited cutinization.

Because these are adaptations to the often conflicting demands of water conservation, gas exchange and light capture, they are governed by the laws of physics (evaporation, gas diffusion, etc.) – laws that we can safely assume to have remained constant while land plants have existed. For this reason, we see many examples of unrelated plants solving the same environmentally-produced dilemmas in similar ways. We can see the same phenomenon in plants from different areas. For example, tropical rainforest vegetation appears to be the same in Africa or South America even though the constituent species are largely different. Desert plants also display such convergent morphologies to the extent that it is often difficult to distinguish between members of the families Cactaceae (cacti which originated in the New World) and Euphorbiaceae (euphorbias which originated in the Old World) without examining their flowers; vegetatively they look identical. We have already used this approach in our considerations of the environmental relationships of Devonian and Carboniferous plants.

Because physiognomic adaptations to environment are so consistent, we can, within certain limits, make quantitative comparisons for determining pre-Quaternary climates. One of the most successful applications of this approach was devised as long ago as 1915, when two American botanists I. W. Bailey and E. W. Sinnott noted that the leaves of modern woody 'broadleaved' flowering plants (such as alder, willow, figs, etc.) tend to have smooth ('entire') leaf margins in warm climates, but toothed, jagged, margins in cool climates (Figure 7.12). Jack Wolfe, another American paleobotanist, developed the methodology further in the late 1970s and, by using modern species growing in drought-free environments in South-East Asia, plotted mean annual temperature (MAT) against the percentage of species with entire margined leaves (see Figure 7.13). It is important to realize that the correlation is less impressive if the growth is limited by water availability in any way – something we will return to in a moment.

This technique of **leaf margin analysis** can be applied to the Alaskan fossils. We know that in mid-Cretaceous time northern Alaska rarely experienced drought, because there are numerous thick coal seams and, although charcoal is present, it is scarce. In addition, tree rings show that growth was uniform throughout the summers and therefore water, to name but one factor, was not limiting. The sediments have yielded a large number of leaf fossils and, to date, 67 different woody broadleaf angiosperms, alone, have been distinguished. Of these, 22 have entire margins.

(a)   (b)

Figure 7.12
Leaf margin types. (a) A variety of tropical rainforest species all displaying smooth (entire) leaf margins. Some also show an extended leaf apex, or 'drip tip', for shedding water. Drip tips are characteristic of plants growing under very wet conditions. (b) A typical temperate tree bearing leaves with toothed margins.

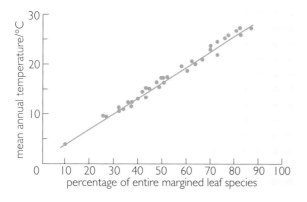

Figure 7.13
Plot of mean annual temperature (MAT) against the percentage of leaf margin types (from Wolfe, 1979). This graph was constructed using plants in drought free environments in South-East Asia. The relationship between MAT and margin type breaks down in dry environments.

Question 7.7

From Figure 7.13, what mean annual temperature (MAT) is indicated by the leaf fossils from the mid-Cretaceous of northern Alaska?

This mean annual temperature is similar to that of London today, but, bearing in mind the polar light regime, it is likely that the mean annual range of temperature

(MART) was much greater. We are making a number of assumptions when we use leaf margin analysis in this instance. First, we are assuming that there is no change in slope of the graph over time, i.e. that the slope of the graph for the plants of today is applicable to the mid-Cretaceous, which was very early in angiosperm evolution. Second, we are assuming that the relationship between margin characteristics and temperature was not different due to any possible effects of the polar light regime. However, Wolfe has found no obvious break in the slope of the relationship above or below latitude 66° (the likely position of the Arctic Circle then), suggesting that the polar light regime had little effect.

Physiognomic analysis can be applied to vegetation types. Here, instead of relying on the response of one group of plants (the woody broadleaves) to climatic variables, all taxa, including gymnosperms (Section 4.6.3), are involved. Because we are using a broad spectrum of taxa, and it is unlikely that they all relate by chance to climate in the same way, this technique is particularly robust. The modern form of this technique is again largely due to the work of Jack Wolfe. Figure 7.14 is a special type of graph called a **nomogram** and each domain (coloured area) represents the temperature field, in relation to mean annual temperature (MAT) and mean annual range of temperature (MART), within which each vegetation type grows. It should now be possible to determine past MAT and MART values if the ancient vegetation can be accurately reconstructed.

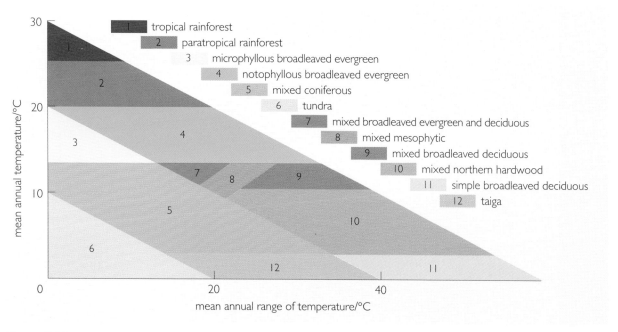

Figure 7.14
Wolfe's nomogram for thermal domains of living forest types. (1) Tropical rainforest. (2) Para (or near) tropical rainforest. (3) Microphyllous broadleaved evergreen forest – this forest type typically has very small leathery leaves and is mostly found in seasonally dry environments. (4) Notophyllous broadleaved evergreen forest – the leaves here are larger than in the microphyllous forest but still leathery and evergreen. Such forests are typically found in areas with a Mediterranean climate. (5) Mixed coniferous – the vegetation is made up of a mixture of conifers such as spruce and pine but with a smaller proportion of broadleaved flowering plants (angiosperms) such as alders and willows. (6) Tundra – vegetation devoid of large trees, found at modern high latitudes and altitudes. (7) Mixed broadleaved evergreen and deciduous – vegetation types that are composed almost entirely of flowering plant trees but with a mixture of evergreens (such as laurels) and deciduous (such as sycamores) species. (8) Mixed mesophytic – these are forests or woodlands made up of a mixture of species with average sized leaves. (9) Mixed broadleaved deciduous – angiosperm-dominated woodlands where most species are deciduous. (10) Mixed northern hardwoods – angiosperm-dominated woodlands where many species are deciduous. (11) Simple broadleaved deciduous – exclusively composed of deciduous angiosperm trees and shrubs. (12) Taiga – a mixture of open shrub, moss and lichen covered landscape and stunted trees such as spruce and larch.

■ Before we do that, however, what is the MAT and MART of tropical rainforest?

■ This occupies a domain between 25 °C and 30 °C for the MAT, and a MART of up to only 10 °C.

Contrast this with taiga (high-latitude conifer woods), which grows at MATs of below 3 °C, but where there is a huge range of temperatures – perhaps exceeding 35 °C.

We do not have the opportunity here to consider in detail how ancient vegetation is reconstructed accurately from dispersed fossilized remains. Suffice it to say that considerable detective work is required to link leaves, wood, and reproductive structures into complete plants of known stature, ecological setting, community association and abundance. This detective work combines detailed observation of plant form with occurrences in a wide range of sediments, in much the same way as we examined the leaf-shedding habits of the plants of the polar forests.

Figure 7.15 shows a reconstruction of the mid-Cretaceous forests of northern Alaska. The dominant plant is a deciduous conifer related to the Bald Cypress and Dawn Redwood. We could characterize this vegetation then as being predominantly composed of conifers. However, there are some other taxa such as river margin angiosperms and ginkgos, with ferns and horsetails as ground-cover plants. This is not a pure conifer forest, then, but is mixed with other components of a broadleaved nature.

Figure 7.15
Reconstruction of northern Alaskan forests of the mid-Cretaceous.

Question 7.8

From the description of the northern Alaskan forest shown in Figure 7.15, use Figure 7.14 to estimate the MAT and MART values of the region.

You may have noticed that the MAT for the northern Alaskan forest is essentially the same as that derived from woody angiosperm leaves, even though in this instance we were using the physiognomy of the vegetation as a whole, including the non-angiosperms. This is an encouraging result as it suggests that our methodologies might be robust.

So far in this section you have seen that it is possible to obtain quantitative estimates of mean annual air temperature by examining the proportion of smooth, as against toothed, woody angiosperm (broad) leaves that occur in a particular vegetation type. You have also seen that leaf size is related to water availability, so by measuring leaf size we have a means by which rainfall might be estimated. In fact, by looking at a range of physiognomic characteristics, we can not only obtain an independent measure of MAT but also some indication of the MART, the average temperature of the coldest month, the mean annual precipitation, the precipitation during the growing season, and even the length of the growing season. The methodology for doing this, again developed by Jack Wolfe, involves some elaborate statistical manipulations which need not concern us here. The important thing to note is that, by using such techniques, the study of ancient vegetation can provide considerable insights into the pattern of past climates.

7.6 Model results

An alternative to the empirical approach to climate reconstruction explored in previous sections is that of theoretical climate modelling, which was introduced in Chapter 5. How do the results of these two methods compare?

Figures 7.16 and 7.17 show the results of a set of simulations run by Paul Valdes of the Meteorology Department of the University of Reading, UK. Using an atmospheric general circulation model derived from that used for medium-term weather forecasting, Valdes has not only predicted, or more strictly speaking retrodicted, the mean annual temperatures and rainfall, etc., but has produced a series of maps representing the average conditions over three-day intervals during an average mid-Cretaceous year. Figure 7.16 represents some three-day average 'snapshots' of the temperatures 2 m above the ground (temperatures dinosaurs, for example, would have experienced), while Figure 7.17 shows the moisture content for the top 10 cm of the soil profile – critical conditions for the establishment of young plants.

If you bear in mind that these retrodictions were not based on geological or paleontological information other than the reconstructed continental configuration and sea-surface temperatures (derived from oxygen isotope data), the results appear to agree, at least to a first approximation, with those derived from the fossil and sedimentological record. In Section 7.3.4 you have seen that plant fossil evidence shows that polar temperatures were considerably warmer than at present. The abundance of forests in near-polar environments argues against permanent ice, and plant architectural features suggest mean annual temperatures of approximately 10 °C for northern Alaska at latitude (at least) 75° N (Section 7.3.1). Although we have no quantitative data for temperatures at low latitudes, the abundance of evaporites (Figure 7.1) and the architectural features of plants such as the *Pseudofrenelopsis* (Section 7.4.1) suggest, on average, low humidities. This could have been caused by either high temperatures or low rainfall, and is likely to have been a result of both.

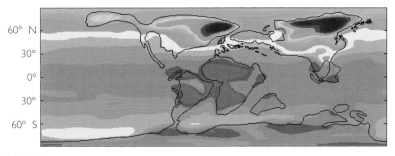

(a) mid-January

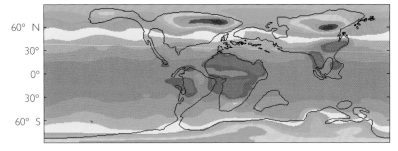

(b) mid-April

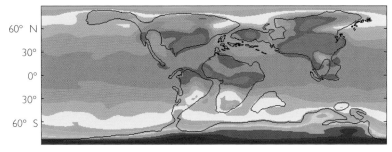

(c) mid-July

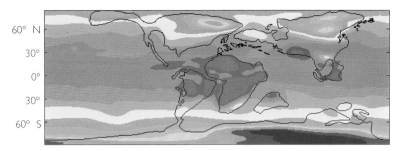

(d) mid-October

<−24 −20 −16 −12 −8 −4 0 4 8 12 16 20 24 28 >32

3-day average temperature 2m above ground/°C

Figure 7.16
Temperature conditions 2 m above the ground for four 3-day snapshots (mid-January, mid-April, mid-July and mid-October) for an 'average year' in the mid-Cretaceous. The maps were produced by an atmospheric general circulation model (AGCM), written by a consortium of British Universities.

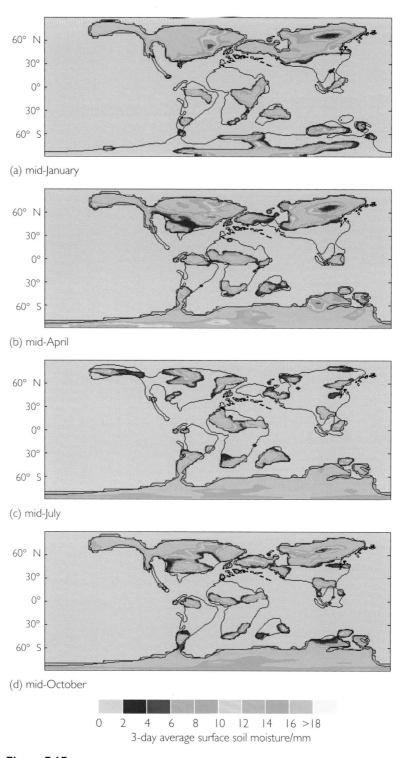

(a) mid-January

(b) mid-April

(c) mid-July

(d) mid-October

0 2 4 6 8 10 12 14 16 >18
3-day average surface soil moisture/mm

Figure 7.17
Soil moisture conditions in the top 10 cm of the soil profile over a 3-day interval in mid-January, mid-April, mid-July and mid-October for an 'average year' in the mid-Cretaceous. These moisture conditions, critical for small plants and seedlings, were predicted using the same atmospheric general circulation model as in Figure 7.16.

Broadly speaking, the model retrodictions are in agreement with the observation noted above. Figure 7.16 shows northern Alaskan coastal temperatures well above freezing for most of the year. Equatorial temperatures, on the other hand, show averages well above 32 °C, which could imply noon temperatures in the order of 50 °C.

The soil moisture maps (examples of which are shown in Figure 7.17) are also in agreement with the geological data. High latitudes are shown to be wet year round, with the top 10 cm of the soil profile only drying for very short periods (less than one week) during the summer. This is in marked contrast with today, where the strong polar high gives rise to very low precipitation.

At low latitudes the model predicts very few regions with an ever-wet climate. Rainfall and therefore soil moisture are spatially and temporally highly variable. Coupled with high temperatures, these factors produce predominantly arid conditions throughout the tropics. The only exceptions to this are in North-West South America and South-East Asia, both of which have onshore ocean winds that supply moisture year round.

▨   What does this imply about the extent of tropical rainforests in the mid-Cretaceous?

▨   It implies that there were very few places that rainforests could develop. Most Cretaceous tropical vegetation was likely to have had characteristics similar to those of modern savannah.

Extensive tropical rainforests are thus a feature of a relatively cool world (though perhaps not *cold* as this tends to dry the atmosphere).

There are some features of the models that require closer examination. You will recall, from Chapter 5, that the models do not have a dynamic ocean; rather they deal only with modelling atmospheric conditions. Sea-surface temperatures based on oxygen isotope data can at least be extrapolated to coastal land areas, but what does the model say about continental interiors where they are less constrained? Well, here there is a degree of controversy. When models have been run for other greenhouse times, specifically the Eocene, the model results predict a far greater annual range of temperatures, and in particular colder winters, than the paleontological evidence suggests. A similar phenomenon exists in the mid-Cretaceous model results in Figure 7.16. The Cretaceous evidence that does exist, again seems to indicate the model temperatures are too low in winter, and rainfall possibly too low in continental interiors.

▨   You may have noticed that nothing has been said about atmospheric concentrations of CO_2 in the Cretaceous models. Why might this be so?

▨   The models were supplied with sea-surface temperatures estimated from Cretaceous $\delta^{18}O$ data. If the models were also given a higher concentration of atmospheric CO_2 than at present, this would result in an enhanced greenhouse effect. We would then be giving the model too much heat by combining (i) the greenhouse heating of higher CO_2 and (ii) the specified higher sea-surface temperatures which ought already to subsume any greenhouse warming. This double-counting would not be appropriate, so the models were run with present-day CO_2 values.

As dynamic ocean–atmosphere models that include vegetation feedbacks are developed, it may be that the disparities between the evidence and the model predictions will be removed. This all goes to emphasize the interconnected nature of Earth Systems even on short time-scales.

7.7 Life and conditions in the sea

7.7.1 Geographical framework

The previous sections concentrated largely upon the Cretaceous vegetation at high latitudes, as this reflected most strikingly the contrast with today's climate. In the marine realm, however, the main differences between then and now were expressed in low to mid-latitudes, both around the Tethys Ocean, which separated the northern and southern continents (Figure 7.18), and in the equatorial Pacific.

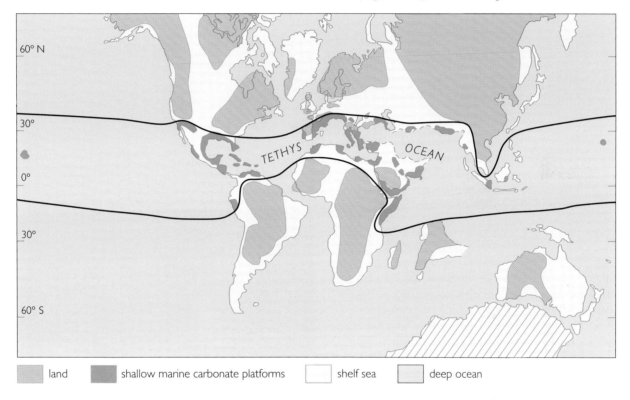

land   shallow marine carbonate platforms   shelf sea   deep ocean

Figure 7.18
The geography of earlier late Cretaceous times, showing the zone of major carbonate platforms formed in and around the Tethys Ocean. The distribution of land and sea in Antarctica is not well known because of the present-day ice cover.

■ Study the map shown in Figure 7.18, and, in particular, those parts of the continents shown lying between the bold lines on the figure (i.e. approximately between the Equator and 30° N). Which is shown as having covered a relatively greater area there, shallow sea or land?

■ Within this paleolatitudinal belt, which also included the Tethys Ocean, shallow sea covered by far the larger proportion of the surrounding areas of continental crust. Indeed, some continental extensions from the North African margin are shown to have formed isolated broad shallow marine platforms, straddling the central part of the ocean.

Elsewhere, seaways spread across the continental interiors – a consequence of the exceptional rise in global sea-level during the Cretaceous Period, already referred to in Section 7.2.

Topographical relief in the land areas that bordered the shelf seas was generally subdued. The main zones of uplift generated by plate tectonic activity around the Tethys Ocean were volcanic island arcs (*The Dynamic Earth*), associated with the subduction of oceanic crust. These lay predominantly along the northern Tethyan margin, where they formed chains of islands and submerged ridges, similar to the Sunda Archipelago of South-East Asia today. The dramatic episodes of low-latitude mountain building, from the Alps to the Himalayas, were to follow later, in the Tertiary, with the collisions between the northern and southern continents that were eventually to close the Tethys Ocean (*The Dynamic Earth*). The mountain ranges that were present in the Cretaceous were situated elsewhere, as for example, in northern Alaska (the Brooks Range: Section 7.3.1).

The climate along the Tethyan belt (i.e. the ocean and its surrounding areas), as you have already seen from the land vegetation (Section 7.4), was predominantly arid, though some uplifted areas appear to have generated local monsoonal systems, with seasonal humidity.

Question 7.9

If you take into consideration the topography and climate of the lands bordering the Tethyan belt, what would you expect to have been the dominant composition of the sediment accumulating in the shallow shelf seas there?

Limestone formed in this way is indeed the most characteristic sedimentary rock type left to us by the shallow seas around the Cretaceous Tethys. Vast tracts of it now form imposing mountains and plateaux, heaved up by subsequent tectonic activity, from Mexico, via the Mediterranean, to the Middle East, and on alongside the Tethyan suture zone of the Himalayas, to South-East Asia (Figure 7.19).

Figure 7.19
A typical massif of (Lower) Cretaceous Tethyan platform limestone, 'Les Calanques', to the East of Marseille, South-East France.

The Cretaceous world thus stood in marked contrast to that of the present day. Today's shelf seas in corresponding latitudes are narrower and mainly limited to the margins of continents. In many instances, mountain ranges shed copious quantities of physically eroded land-derived sediments into them. These sediments both inhibit the growth of many carbonate-sediment producing organisms (e.g. corals and calcareous algae) and swamp any shell debris that is produced. Such conditions reflect both today's relatively lower global sea-level, and the extensive mountain building that has occurred in low and mid-latitudes since the Cretaceous. Consequently, shallow tropical carbonate provinces are now much less widespread than those of the Cretaceous.

7.7.2 Tethyan carbonate platforms

Limestones are a crucial component of the carbon cycle, and hence, as you should by now realize, of the climate system. The limestones formed around the Cretaceous Tethys are therefore worth a closer look, firstly to see how they came to be deposited so copiously, and, secondly, in order to estimate their quantitative contribution to carbon burial during the period.

Characteristically, the Cretaceous Tethyan limestones accumulated so as to form broad, shallow, submarine plateaux, termed **carbonate platforms**, the outer margins of which sloped away into neighbouring basins. The depth of the basins varied greatly, according to tectonic setting (i.e. tectonic regime of a region), but within most areas of continental crust it usually ranged from only some tens to a hundred or more metres.

In recent years very detailed studies have been carried out on the geometry and constitution of the beds composing such platforms, with the aim of reconstructing how the platforms grew. An example is illustrated in Figure 7.20, which shows a vertical section across the edge of the Vercors platform, of early Cretaceous age, in eastern France. It shows a pattern seen very frequently in Cretaceous Tethyan carbonate platforms: sets of inclined beds of relatively coarser sediments (mainly shell debris), derived from the top of the platform, project obliquely down like overlapping tongues into the finer sediments of the basin. Thus the platform margin gradually built out laterally into the basin (**progradation**), as the coarser sediment spilt down its flanks. The transport of this material off the top of the platform was a consequence of more shell debris being produced there than could be accommodated in the shallow waters covering the platform: in other words there was an over-production of sediment by the platform 'carbonate factory', which was swept off by waves, tides or storm currents onto its flanks. Between these periods of progradation, occasional subsidence of the platform, or rise in sea-level, allowed basinal sediments to creep up the flanks again until such time as progradation was next resumed. These episodes yielded the thin wedges of finer material that separate the progradational tongues.

Figure 7.20

Diagrammatic section across the platform to basin transition of the Vercors platform in eastern France, showing the geometry of the constituent beds, based on a synthesis by Arnout Everts.

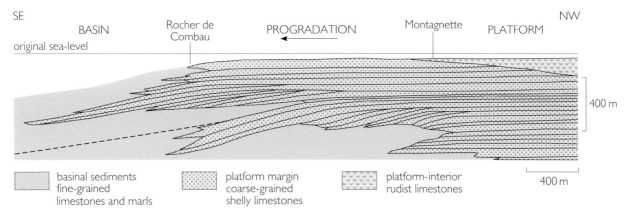

There was, inevitably, much variation on this basic depositional theme, as a result of differences in the rates of deepening (due either to subsidence or eustacy) and in the multitude of factors affecting sediment production and transport on the platforms. Thus, for example, where contemporaneous faulting at the platform margin accentuated the depth of the neighbouring basin, the sediment spilling off the platform had a larger space to fill, so marginal slopes became steeper and outward growth less marked. On the other hand, the platforms themselves sometimes became 'drowned', when their carbonate factories failed to keep up with sea-level, either because of adverse conditions in the water overlying them, or because of too-rapid deepening, or a combination of both.

Nevertheless, for much of their development the platforms tended to build up and outwards in fits and starts, as space to accommodate the sediment they produced was provided by successive deepening events. At this point you would be justified in asking 'why so much repeated deepening, and why were there not just as many episodes of emergence, leading to weathering and dissolution of the limestones?' Here, we come to a crucial distinction between the Cretaceous greenhouse world, in which these platforms developed, and today's icehouse world.

Question 7.10

Which major cause of occasional rapid eustatic falls in global sea-level in the Quaternary would have been absent from the Cretaceous world?

There is some evidence (e.g. marine dropstones) suggestive of a few minor glacial advances to sea-level in the early Cretaceous. But even this evidence is disputed, with critics claiming that the boulders could have been rafted on floating tree trunks. Any brief glaciation there might have been was certainly never on a scale to compare with those of the Quaternary.

By contrast, glacially-driven sea-level change has been instrumental in shaping the tropical carbonate platforms of today, giving them a geometry unlike that of the Cretaceous platforms. Many of the modern platforms are rimmed by sturdy coral reefs (Figure 7.21 and see the frontispiece of this book), which reach up to sea-level and form prominent barriers to incoming ocean waves.

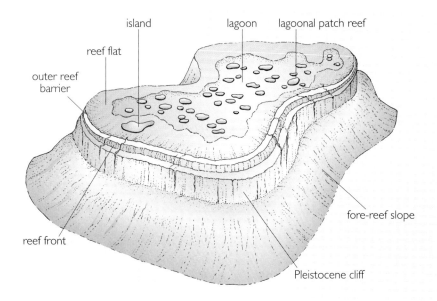

Figure 7.21
Cartoon of a typical modern carbonate platform rimmed by a barrier reef.

It would be tempting to suppose that the modern barrier reef topography, such as that shown in Figure 7.21, was entirely created by the vigorous upward growth of the coral reef itself at the platform margin. In many cases, however, the reef itself is installed on a pre-existing topography that already had a raised margin (Figure 7.22). How might that have got there? The glacio-eustacy mentioned above provides the answer to this question. Commonly, the reefs have grown upon yet older carbonate platforms, which had been emergent during the last glacio-eustatic fall. Dissolutional weathering of these older platforms often left behind projecting edifices around their edges, rather like castle walls, cut across here and there by gullies. When the weathered platforms later became submerged again, these prominences (the Pleistocene cliffs in Figure 7.21) provided the hard foundations for the re-establishment of reefs around the margins, as shown in Figure 7.22.

The Cretaceous world, as noted above, lacked such glacio-eustatic fluctuations of sea-level, at least on any comparable scale. Hence significant emergence and weathering of the carbonate platforms was infrequent, except where local tectonic activity intervened. Instead, the platforms tended to experience only successive minor deepening events brought about through a combination of the overall eustatic rise of the period, and regional subsidence of the crust. Between these sporadic deepening events, carbonate sediment over-production led to the cycles of progradation discussed above.

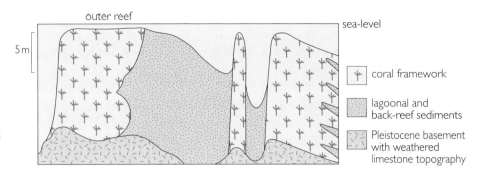

Figure 7.22
Diagrammatic cross-section across the margin of One Tree Reef, southern Great Barrier Reef, Australia, showing the siting of the reefs upon pre-existing topographical prominences.

How did this characteristic pattern of growth of the Cretaceous platforms affect the organisms that dwelt on them, and ultimately generated them? The common lack of inherited topographical relief sculpted by emergence and weathering, and the frequent fluxes of carbonate sediment across the prograding margins, meant that there was little opportunity for the stable establishment there of barrier reefs. The platforms thus lacked the sturdy protective rims created by barrier reefs today, which take the brunt of incoming waves from the open sea and limit the flux of water and sediment across the platform margins. Instead, the outer parts of the Cretaceous platforms tended to be dominated over broad areas by migrating banks, or sheets, of current-swept shell sand and debris, which graded into expanses of muddier lime sediments in the platform interiors. These broad sediment surfaces supported myriads of bottom-dwelling shelly organisms, the growth and death of which in turn fed back more sediment to the system. Foremost among the larger skeletal sediment-producers, particularly in the late Cretaceous, were some oddly shaped gregarious bivalves called rudists (Figure 7.23). These grew implanted in, or lying upon, the loose sediment, as vast 'meadows' of clustered shells, rather like oyster beds today. Frequent disruption of these congregations, usually by storm currents, together with breakage and perforation of the shells by burrowing and boring organisms, helped to mill them into shell sand and mud. The upward and outward growth of the platforms was thus self-sustaining, as shell debris was harvested over huge areas and spread both inwards, and out onto the marginal slopes.

Sometimes these rudist-dominated formations are referred to in textbooks as 'rudist reefs', but, as the description above shows, this is quite a misleading term as far as the anatomy and growth of the carbonate platforms were concerned. Rather, the rudists were essentially level-bottom sediment-dwellers, forming shelly 'meadows' and associated carpets of redistributed debris.

Figure 7.23
Clusters of tubular rudist bivalve shells, preserved mostly in life-position implanted in the sediment in which they grew, from the Upper Cretaceous of (a) Provence, South-East France, and (b) the Spanish Pyrenees.

(a)   ~20 cm

(b)   ~10 cm

7.7.3 Quantifying the growth of the platforms

Let us now consider some quantitative aspects of these Cretaceous Tethyan carbonate platforms. Their areal extent naturally varied quite a bit through the Cretaceous, with long periods of expansion being punctuated by brief episodes of decline, when carbonate production was temporarily reduced or even halted on many platforms because of adverse conditions. Following the latter, large tracts of the platforms commonly continued to subside, and so they became 'drowned' in deeper water. This in turn necessitated their regrowth by progradation from shallower areas when conditions improved again. A major study by a team of French geologists has investigated quantitative aspects of their history. One of the main phases of platform growth was approximately halfway through the Cretaceous, between about 94 Ma and 92 Ma ago (contemporaneous, in fact, with some of the northern Alaskan forests discussed in Section 7.3). The Tethyan carbonate platforms at that time have been estimated to have covered a total area of some $9.68 \times 10^6 \, \text{km}^2$ (an area approximately equivalent to that of China or Europe). Their thicknesses, of course, varied according to the amount of accommodation space provided for their accumulation by subsidence, itself dependent on the tectonic setting. However, values from several examples of this age from around the Middle East and the Mediterranean commonly range between 10 and 100 m per million years, with most closer to the lower end of the range. A value of 30 m per million years is a reasonable estimate of their average vertical rate of accumulation for the purposes of calculating burial rates.

> **Question 7.11**
> From the figures above, calculate the average rate of burial of carbon in these platform limestones, in kilograms per year, given the following further information: the limestone may be regarded as pure calcium carbonate ($CaCO_3$), with a bulk density of $2.71 \times 10^3 \, \text{kg m}^{-3}$; and the proportion of carbon (C) in limestone accounts for approximately 0.12 of its mass.

Compare the answer to Question 7.11 with the estimated total rate of burial of carbon in marine carbonate sediments – of all kinds (including oceanic oozes as well as shallow carbonate sediments) – today, which is $0.2 \times 10^{12} \, \text{kgC yr}^{-1}$ (*The Dynamic Earth*). The astonishing conclusion is that the Tethyan carbonate platforms *alone* at this time were burying carbon (in the form of carbonate) at nearly half the total rate for *all* marine carbonate sediments today. To put this figure in context, the estimated rate of carbon burial in shelf limestones (i.e. other than on the ocean floor), at *all* latitudes today, is about $0.07 \times 10^{12} \, \text{kgC yr}^{-1}$.

7.7.4 Other sinks for carbon

These Tethyan platforms were by no means the only sites of carbonate deposition in the Cretaceous seas. The basins neighbouring them, on subsiding areas of continental crust around the Tethys Ocean, frequently accumulated fine-grained carbonate sediment, or mixtures of carbonate mud and clay (marl). The volumes involved were also considerable, though less easily calculated.

At mid-latitudes, too, vast areas of the continents were blanketed by deposits of **chalk**, a distinctively fine-grained kind of limestone. The white cliffs of Dover are probably the best-known example of such deposits, but similar thicknesses of chalk of late Cretaceous age may be traced across much of northern France and the low countries, into Germany, Poland and on into Russia, while similar deposits are also present across large tracts of the American mid-West as well as in Western

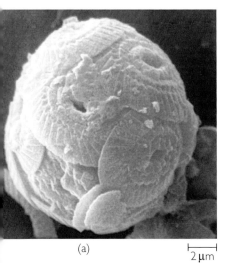

(a)

2 μm

(b)

2 μm

Figure 7.24
Scanning electron photomicrographs of coccoliths of late Cretaceous age: (a) shows a complete cell coating (coccosphere) of coccolith plates; (b) shows an isolated coccolith, from another species.

Australia. Deposited at slightly greater depths than the Tethyan platform limestones (mainly between 50 and 200 m), these were the Cretaceous equivalents of the calcareous deep-sea 'oozes' of today. Because of the high global sea-levels of the time, however, much of the chalk ooze was then deposited over drowned continental areas, rather than being limited to the shallower ocean floor, as is the case today. Chalk is composed mainly of the minute skeletal plates of some planktonic unicellular algae called coccolithophores (Figure 7.24), accompanied by the remains of other calcareous planktonic forms, such as certain kinds of foraminifers (protists), as well as those of various bottom-dwelling organisms. Although calcareous-shelled plankton had first evolved some time earlier (at least by the late Triassic), their abundance increased dramatically in the late Cretaceous, yielding the extensive thick chalk deposits mentioned above.

Calculating the rate of accumulation of the Cretaceous chalks is a little harder, because of variations in their thicknesses and compositions. However, at a crude approximation the European chalks alone, for example, probably buried carbon at a rate approximately equivalent to the Tethyan carbonate platforms. Still more carbonate was deposited in the oceanic realm.

Question 7.12
In *The Dynamic Earth*, you encountered the concept of the 'calcium carbonate compensation depth' – the depth in the ocean water column at which dissolution of $CaCO_3$ (as a consequence of pressure, temperature, chemistry, etc.) exceeds supply from the surface waters; below this depth, $CaCO_3$ does not accumulate on the ocean floor. What effect do you suppose the high sea-levels of late Cretaceous times would have had on the accumulation of *oceanic* carbonate sediments?

On the other hand, from the admittedly crude estimates discussed above, it would appear that total rates of carbonate accumulation on continental crust in the Cretaceous (i.e. in the Tethyan carbonate platforms, their contiguous basins and the chalk shelf seas) already surpassed the *total* carbonate burial rates of today. *Any* oceanic component would have merely added to the relative excess.

What is clear, then, is that carbon was being buried in Cretaceous limestones at a rate significantly greater than that of today. Exactly by how much is not yet known, because precise estimates have still to be calculated.

Nor can any compensating decline in other carbon sinks be readily detected. If anything, they too seem to have increased their appetites! In addition to the prolific burial of coal on land in high latitudes, as discussed in Section 7.3, huge amounts of organic carbon were also buried in the sea. A combination of factors allowed the latter to happen. In contrast to the cold saline bottom waters of today's oceans (*The Dynamic Earth*), the Cretaceous oceans tended to acquire deep reservoirs of warm saline water produced by intense evaporation at the surface (especially in the arid low latitudes). While the cold bottom waters of today are relatively well oxygenated, their warm counterparts in the Cretaceous were much less so. There were several reasons for this, one being simply the declining solubility of gases as water temperature rises. From time to time extensive water masses within the Cretaceous oceans even became anoxic, allowing abundant organic material to rain down, without being oxidized on the way, and to accumulate in the bottom sediments. Moreover, the complex geometry of the Tethys Ocean and its surrounding seas (Figure 7.18) created numerous more or less restricted basins, the bottom waters of which were particularly prone to stagnation and anoxia. Indeed, many of the giant oil fields of the Arabian Peninsula contain oil from Cretaceous source rocks that accumulated in such basinal areas within the continent, and which migrated to, and became trapped in, platform limestones that flanked the basins.

Question 7.13
From what you read earlier in this book, how might you expect episodes of enhanced organic burial to be reflected in the isotopic ratio of carbonate carbon in marine limestones?

A detailed analysis of $\delta^{13}C_{carb}$ values in the English Chalk, by Hugh Jenkyns at Oxford University, and his colleagues, for example, has shown a strong positive shift in values within Upper Cretaceous strata corresponding to the end of the mid-Cretaceous episode of carbonate platform building that was discussed in Section 7.7.3 (Figure 7.25). A similar change in Italian sequences of the same age suggests a widespread influence then from briefly enhanced rates of organic carbon burial.

From the peak values of $\delta^{13}C_{carb}$ for the Cretaceous, it has been estimated that the rate of marine burial of organic carbon may at times have reached up to three times its present value. However, the timing of these episodes of organic burial in relation to the growth of the carbonate platforms is by no means simple, and is still not well resolved. There may have been a degree of counter-balancing, with one increasing as the other declined (as noted above in relation to Figure 7.25). However, it is also evident that, at times, both organic burial and carbonate platform growth were occurring synchronously, though in different regions.

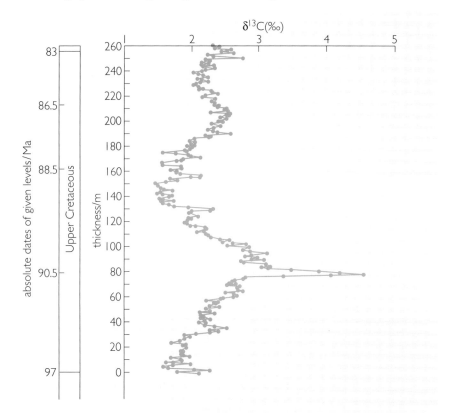

Figure 7.25
Values of $\delta^{13}C_{carb}$ in the chalk sedimentary succession of East Kent recorded by Hugh Jenkyns and colleagues. Note that the vertical scale corresponds to thickness of the sedimentary succession from the base of the chalk. The time-scale is therefore irregular as rates of deposition varied. The dating of certain levels is nevertheless also indicated. The distinct peak at around 80 m corresponds to the end of the mid-Cretaceous episode of carbonate platform building discussed in Section 7.7.3. .

Question 7.14

What, meanwhile, would have been the effect of these two forms of carbon burial upon the supply of molecular oxygen to the atmosphere?

Notwithstanding the need for many details of timing still to be worked out, it is already clear that the overall rate of burial of carbon in the Cretaceous greatly exceeded that of today. Where all the extra carbon may have come from poses an interesting problem. Such a high rate of burial of carbon could not possibly have been sustained throughout the period on credit, so to speak, from the Earth's surface reservoirs of carbon alone, without compensation from elsewhere. The atmospheric reservoir (taking today's figure of 0.76×10^{15} kg of carbon) would have been exhausted in a matter of thousands of years: clearly it wasn't, as the loss of this greenhouse gas would have caused a drop in global temperature, in contrast to the notably warm climatic conditions of the period. Even the total reservoir of carbon in the soil and oceans would have lasted perhaps only some tens of thousands of years: that, too, was not in fact detectably depleted, as there is no shortage of fossil evidence for the healthy continuation of life through the Cretaceous. Hence we must look to the Earth itself for the extra carbon.

7.8 A surfeit of carbon: the key to the Cretaceous greenhouse

7.8.1 *De profundis*

We must now consider how such an excessive supply of CO_2 could have been sustained.

Question 7.15

If you look again at the paleogeography of the Cretaceous world (Figure 7.18), do you suppose that the extra carbon could have been furnished either by increased weathering of fossil fuels, or as a consequence of decreased weathering of silicates (a process that is an alternative sink for CO_2, as explained in Section 6.3.2)?

The only remaining large-scale source of carbon is volcanic outgassing of CO_2. As discussed in *The Dynamic Earth*, there is widespread evidence in the western Pacific, in particular, for massive volcanism in the Cretaceous, probably associated with a mantle superplume. Unlike rift volcanism, the volcanoes associated with this intraplate source reached to sea-level, where their gases would have been released directly into the atmosphere.

Such volcanic activity in the Pacific seems to have been spread over a long time interval, largely between about 125 Ma and 80 Ma ago, with a major initial phase in the first 15 Ma, followed by a later phase peaking during the last 10 Ma. The calculations of carbon burial rates you did earlier for the Tethyan carbonate platforms concentrated on just one part of the mid-Cretaceous (around the start of the second maximum of volcanism), by way of illustration. However, such carbonate platforms developed extensively, though with episodic drowning, throughout the interval in question. The anomalously high sea-levels of the time, particularly of the late Cretaceous, can also be linked with the widespread thermal doming of the ocean floor associated with the volcanism.

7.8.2 The greenhouse atmosphere

If we now have a plausible explanation for the hyperactivity of the Cretaceous global carbon cycle, as well as the elevated sea-levels of the period, in the form of the 'Pacific superplume', one vital piece of the jigsaw remains to be put in place: what effect did the enhanced flux of carbon have on atmospheric levels of CO_2, and hence on climate? This is a tricky question, because we cannot assume a simple connection between the rate of through-flow of carbon in the atmosphere, and its level in that temporary reservoir. Yet it is the latter that would have exerted an influence on climate through its greenhouse effect. Think of the analogy of running water from a tap into a bucket that has some holes in the bottom. Given a sufficient rate of supply, the water in the bucket is going to fill up to a certain level at which the head of water in the bucket causes the rate of outflow to match that of the inflow. Thereupon the water level will remain balanced. Increase the rate of inflow, and the level will of course rise to a higher level. But increase the size of the holes, so increasing the outflow, and the water will drop back to a lower level again if the same rate of inflow is maintained. The rate of through-flow would then have increased without necessarily raising the eventual water level. However, the fluctuations in level in this analogy suggest a possible model for the Cretaceous carbon cycle.

In the Cretaceous world, we may consider the volcanoes as equivalent to the tap in the analogy, ultimately supplying the carbon. Changes in volcanic activity would have caused fluctuations in the rate of supply. Continuing the analogy, the Tethyan carbonate platforms, say, may be regarded as one of the larger 'holes' for the outflow of carbon. The overall surface area of the platforms then corresponds to the size of the 'hole'. As we have seen above, the areal extent of the platforms did indeed fluctuate quite markedly with time, with short episodes of widespread platform drowning punctuating longer periods of re-establishment and progradational expansion. This pattern implies a degree of necessary lag between any increases in carbon supply (from volcanic CO_2 emissions) and compensatory expansion of the platforms, serving to 'soak it up'.

> **Question 7.16**
> What would you expect to have happened to CO_2 levels in the atmosphere during these lag intervals?

What about the reverse – a decrease in volcanic emissions? Reversing the argument above, we could expect CO_2 levels in the atmosphere to have dropped, as the giant platforms continued to draw down carbon. The ensuing climatic cooling (with associated effects) is likely to have been deleterious to the growth of the shelly organisms on the carbonate platforms. If carbonate production on the platforms then failed to keep up with their subsidence, the platforms would have rapidly become drowned so reducing the sink for carbonate carbon and thus allowing atmospheric levels of CO_2 to start climbing again. Oxygen isotopic studies do indeed suggest that some of the major episodes of platform extinction and consequential drowning were associated with climatic cooling. In some instances, moreover, the cooling itself appears to have been linked with enhanced rates of organic carbon burial (as in the example shown in Figure 7.25). When volcanic emissions picked up again, there would again have been under-compensation from the platforms, and a new cycle of platform growth would have ensued.

Thus the lag between volcanic emissions and platform growth – a mismatch between the rates of change of the fluxes supplying and burying carbon – could have led to increased levels of atmospheric CO_2 throughout much of the Cretaceous, as suggested by the GEOCARB model mentioned earlier (Chapter 6). Such increased levels would certainly have contributed to climatic warming. The ecosystems that arose in response – such as the Tethyan carbonate platform communities and the polar forests discussed in this chapter – reflect the collective adaptive responses of organisms to the conditions engendered. They have no equivalents in today's icehouse world, which has its own, equally distinctive, ecosystems.

7.9 Conclusions from the case studies

Both the greenhouse world of the Cretaceous, and for that matter the icehouse world of the Permo-Carboniferous (discussed in Chapter 6), reflect the complex process of continuous re-equilibration between the feedbacks of evolving life and the changing Earth. Conditions at the Earth's surface have clearly fluctuated, primarily in response to changes within the Earth itself (controlling paleogeographical organization and superplume activity, for example) but also as a consequence of evolutionary innovations with major environmental impacts (largely upon atmospheric composition, but also upon weathering rates and sedimentation, for example). Other influences have included changes in solar radiation and occasional impact-related perturbations. No consistent stable state is evident, each age finding its own unique balance of interacting influences.

7.10 Summary of Chapter 7

1   The Equator-to-pole temperature gradient was much shallower in the Cretaceous than the present – the difference being largely due to much warmer poles.

2   Low latitudes were seasonally arid and, compared to today, there were few areas where rainforest could develop. Instead, the low-latitude plants display special adaptations to conserve water.

3   The polar light regime was similar to that of the present with prolonged periods of winter darkness.

4   The Arctic was devoid of permanent ice and supported luxuriant forests dominated by deciduous conifers, but with ferns, ginkgos, cycads and some angiosperms also present. These forests were effective carbon-sequestering systems.

5   Delta flood plain accumulations of peat were buried as a result of subsidence and sediment shedding from nearby mountains.

6   Paleoclimatic determinations cannot be based reliably on relictual species such as the cycads whose modern representatives only poorly reflect their past biological diversity and climatic tolerances. Instead, more time-stable techniques based on physiognomy have to be used.

7   Analysis of the physiognomy of the total vegetation, as well as that of the leaves of woody 'broadleaved' flowering plants, can yield quantitative climate data. Both techniques provide similar results suggesting they are reliable. Such studies have shown that mean annual temperatures at between 75 and 85° N were much warmer than now at such high latitudes.

8   Paleoclimate can be studied using atmospheric general circulation models. Model results are generally in good agreement with data obtained from the geological record.

9   Flooding of the continents provided broad shallow seas around the equatorial Tethys Ocean, flanked by predominantly arid lands of low relief. These seas proved favourable sites for the development of extensive carbonate platforms.

10   The carbonate platforms frequently expanded laterally (prograded) through excess carbonate production being swept off the tops of the platforms, onto their flanks. Episodically, basinal sediments crept back up the flanks, with brief phases of deepening.

11   The Cretaceous platforms differ in structure from tropical platforms today because of the effects only on the latter of significant glacio-eustatic fluctuations. Unlike today's platforms, the Cretaceous examples typically lacked marginal reefs, and their outer zones were instead dominated by migrating banks of current-swept shell sand and debris. The prolific growth of shelly organisms on these surfaces fuelled the massive carbonate sediment production of the platforms.

12   The rate of burial of carbonate carbon on the Tethyan platforms alone at times approximated to half that in all carbonate sediments – of deep, and shallow, water origin – today. However, to the Cretaceous total must also be added that buried in Tethyan basins and mid-latitude chalks as well as any oceanic deposits. Given the high rates of burial of organic carbon at that time, too, carbon was being buried in Cretaceous sediments at a significantly higher rate than it is today.

13   The only likely source for the excess carbon in the Cretaceous is increased volcanism, most plausibly associated with the 'Pacific superplume'. Rising levels of CO_2 in the Cretaceous atmosphere can be attributed to lags between increases in volcanic emissions and compensatory growth in the carbonate platforms. Episodic extinction and drowning of the latter may have been associated with climatic cooling brought about by temporary net excesses of CO_2 drawdown.

14   The icehouse and greenhouse case studies of Chapters 6 and 7 reflect the complex process of continuous re-equilibration between the feedbacks of evolving life and the changing Earth. No consistent stable state is evident, each age finding its own unique balance of interacting influences

Chapter 8
The relationship between evolving life and the Earth

8.1 Introduction

We can now return to the question that was posed at the beginning of this book: what is the fundamental nature of the relationship between evolving life and the Earth – benign partnership, or a chaotic system lurching from one temporary state of balance to another?

8.2 Possible worlds

Before considering the evidence, let us briefly rehearse the range of possible answers to this question. At one extreme, one might postulate that the role of life had been purely passive, with no significant effect on conditions at the Earth's surface: organisms have merely adapted to changes dictated by the Earth's physical and chemical state. At the other extreme, the Earth, together with the life it supports, might be seen as a kind of 'superorganism', which has evolved a tightly coupled system of feedbacks ensuring that 'the Earth's surface environment is, and has been, regulated at a state tolerable for the biota': this is the 'Gaia hypothesis' of James Lovelock (Box 1.3). Between these two extremes is the view that, while life has significantly altered conditions at the Earth's surface relative to what a lifeless Earth would have been like, the feedbacks concerned are not tightly coupled in any manner akin to the homeostatic mechanisms of an individual organism: at most, they may contribute, along with the Earth's abiotic processes, to temporary equilibria. Nevertheless, these equilibria are not stable over the long term, being subject to alteration due to changes in the balance of feedbacks both from the Earth and from evolving life itself, including the effects of sporadic major environmental perturbations.

Before going on, you may wish to ponder these three possibilities in the light of what you have read, to see which you consider the most plausible view. Consider, first, whether or not the presence of life has had any significant effect on conditions at the Earth's surface. If so, consider next whether or not such influences have consistently tended to regulate conditions in a state 'tolerable for the biota'.

8.3 Review of the options

Hopefully, you will not have devoted too much time considering the first option – life as a purely passive passenger on Earth's voyage through time. The scale of life's impact on the Earth has been repeatedly stressed throughout this and the preceding books of this Course.

Question 8.1
Think of one major aspect of the Earth today that would have been utterly different had the planet remained lifeless.

That leaves the nature of such feedbacks between life and the Earth to consider – Gaian, or merely contributors to complex, but essentially chaotic, biogeochemical cycles?

Consider, first, the expectations of the theory of evolution by natural selection, discussed in Chapter 1.

Question 8.2
From this perspective, can we expect feedbacks between Earth and life to have generated a tightly coupled, truly homeostatic, system?

If natural selection theory thus provides no grounds for the Gaia hypothesis, what, then, of the empirical evidence? Does the geological record support the proposition that 'the Earth's surface environment is, and has been, regulated at a state tolerable for the biota'? Might such self-regulation have come about by means other than evolution by natural selection? The first problem in tackling this question is the woolliness of the proposition itself. Which 'biota' is being talked about? You have seen that the Earth's biota has changed quite considerably over time. Were you able to question (and to get an answer from!), say, the anaerobic denizens of the Archean oceans, to whom today's levels of molecular oxygen would have been toxic, or the high-latitude forests of the Cretaceous Period, whose domains are now icy wastes, on the issue, their responses would not have been wholeheartedly supportive of the Gaia hypothesis. Nor, apparently, were these life-forms the unfortunate victims of mass extinctions, which might be blamed on extraneous perturbations: their environments simply changed beyond their limits of tolerance and they were eventually written out of the drama (or, in the case of the anaerobic microbes, literally went underground, or into the guts of animals including ourselves, where their descendants continue to thrive in anoxic regimes). Obviously, then, the reference to the biota in the proposition is not intended to be comprehensive: presumably *any* manifestation of life is intended.

The observation that life in general appears to have persisted at least from early Archean times is trivial, in this respect, as it is merely consistent with, but does not necessarily confirm, the Gaian hypothesis: it does not disprove the alternative model that life has managed to continue participating in a chaotic system that lurches between temporarily equilibrating states. It could simply be the case that the 'lurching' has not, so far, transgressed the limits of tolerance of all life-forms: some form of life has always been able to survive, furnishing successive ages with appropriately adapted organisms. In other words, there has always been somewhere, so far, where DNA has proved robust.

So, does the geological record suggest any *predominant* tendency towards the establishment and stabilization of optimal conditions? Demonstration of the mere existence of some feedbacks having this effect would not be enough to settle the issue, as some such effects could be expected in the chaotic model – by chance alone. The Gaian model explicitly proposes a coherent integration of feedbacks, together yielding homeostatic self-regulation. A suitable test, therefore, is to see whether such a pattern has been predominant over geological time. If it has not – with the feedbacks being sometimes stabilizing, sometimes not so – then the Gaia hypothesis should be rejected.

Question 8.3
Consider, for example, the various feedbacks involved in regulating Permo-Carboniferous atmospheric composition and hence climates, which were discussed in Chapter 6. Do these appear to have consistently stabilized a particular set of conditions? Or were their effects variable – sometimes maintaining a given state, and sometimes helping to install a different regime?

Again, in the Cretaceous (Chapter 7), the extensive carbonate platform biota, as well as the high-latitude land flora, did not enjoy uninterrupted exploitation of the enhanced supply of atmospheric CO_2 and the associated warm climatic conditions. It seems that they were poorly suited to respond to hiccups in the supply, as they continued to draw down carbon at excessive rates, and so perhaps contributed to episodic crises involving climatic cooling and eventual drowning of the platforms themselves (Section 7.8.2). Thus, far from consistently stabilizing optimal conditions for the incumbent biota, the Cretaceous Earth–life system became implicated in important environmental fluctuations, which were attended by extinctions on a variety of scales. The same story emerges for any geological period when investigated in sufficient detail.

Of the two models that we have been considering, then, that of a complex, essentially chaotic, system seems to be the more plausible. The Gaia hypothesis is supported neither by evolutionary theory nor by the empirical evidence of the geological record. The real world, it seems, does not always behave like 'Daisyworld' (Box 1.3).

That having been said, it is worth pondering how the fluctuations in conditions remained within the limits of tolerance of living organisms, and why the conditions that we see around us today seem to be so well suited to the living biota. It is, in fact, easier to answer the second question first, as it is really a trick question, but tackling it helps a little way towards answering the first, and it also raises the crucial issue of rates of change.

While the 'rapid' fluxes of biogeochemical cycles that are largely mediated by ecological interactions may balance out over the short term (as explained in *The Dynamic Earth*), slower geological and evolutionary processes alter such equilibrium states over longer time-scales. Atmospheric oxygen is a case in point, the nice mutual adjustment of major sources and sinks (e.g. photosynthesis, respiration, etc.) over the time-scales of human observation providing a telling contrast to the dramatic long-term changes of the past that were documented in *Atmosphere, Earth and Life*. So long as these slower changes do not outstrip the limits of tolerance conferred on individuals by available genes, evolution can deliver suitably adapted organisms. Thus, to marvel at the appropriateness of conditions for life is simply to put the cart before the horse: it is the evolutionary responsiveness of organisms that has ensured the fit. With a little bit of adaptation, today's problem can even become tomorrow's manna. This is known as the 'helpful stress effect'. In recent times, rats have provided a good example of such a change of status, in their evolution of resistance to the pesticide Warfarin. Warfarin is poisonous to normal rats because it suppresses blood-clotting (leading to fatal bleeding) and interferes with the uptake of vitamin K. Soon after its introduction, rats resistant to the pesticide began to appear and to replace the ordinary Warfarin-sensitive strains. However, the frequency of the resistant forms rapidly declines when use of the pesticide is stopped. Thus Warfarin-resistance evidently bears a cost to fitness under normal circumstances (probably related to an increased vitamin K requirement), such that the Warfarin-resistant strains positively depend on the presence of the pesticide in order to survive.

Question 8.4
How does the inferred pattern of evolution of the earliest eukaryotes illustrate this principle in relation to the appearance of molecular oxygen?

Stark testimony to the importance of this principle in the history of life is provided by instances where rates of environmental change have evidently outstripped the capacity for evolutionary response of organisms: the results have been dramatic – extinction, and in cases where such perturbations were widespread and pervasive in their effects, mass extinction. Thus the *rate* of an environmental change, perhaps rather more than its nature, is crucial to its effect: be it slow enough, and it provides new opportunities for adaptation; be it too fast, and it becomes a catastrophe, leaving a trail of extinctions in its wake. Emissions of CO_2 associated with major volcanic episodes seem to illustrate the contrast between these two effects. Brief but massive eruptions of flood basalts may well have been implicated in some mass extinctions (Section 6.6.2). Yet those associated with the postulated Pacific superplume of the Cretaceous, which were erupted on an even greater volumetric scale, but over a longer period, helped to create the conditions in which the spectacular carbonate platform-, and high latitude forest-, ecosystems of the period thrived (Section 7.8).

The other question raised above was: how did the fluctuations in conditions remain within the limits of tolerance of living organisms? To some extent, the progressive evolutionary tracking of changing conditions discussed above can help to explain this. The limits of tolerance have themselves shifted in some respects. Had the physical and chemical conditions of today's world been miraculously switched on in, say, the Archean or the early Proterozoic, the effect on life then would certainly have been devastating, perhaps even fatal (not unlike that of a huge dose of bleach!). Nevertheless, that is certainly not the whole story: the limits within which mean global surface temperatures appear to have remained, for example, are impressively modest. It is a curious but characteristic feature of complex systems, involving myriads of feedbacks between component parts, that they tend to settle within relatively narrow limits of physical and chemical states. The numerical modelling of such behaviour involves a difficult branch of mathematics – chaos theory – that is beyond the scope of this Course. An important point to note, however, is that such sets of conditions ('strange attractors', as they are known) are far from inviolable. As you have already seen, they lack the protection of co-adapted homeostatic mechanisms characteristic of individual organisms. They are indeed quite sensitive to initial conditions, and unpredictable shifts to new states can result from small but critical changes in the components of the system. Possibly, then, mean global temperature may have been broadly constrained in this manner, though evidently it fluctuated between the contrasting greenhouse and icehouse modes we have discussed. Perhaps, moreover, we have just been lucky, in that the early removal of carbon from the Earth, with the formation of the Moon (*Origins of Earth and Life*), followed by the sustained drawdown of CO_2 from the atmosphere, removed the risk of a runaway greenhouse effect as the Sun warmed. We do not have a satisfactory answer to this question at present.

8.4 Conclusion

In summary, it is clear that evolving life has profoundly influenced conditions at the Earth's surface over time. Yet it is equally apparent that the feedbacks between Earth and life have interacted over the long term in a chaotic, undirected, manner, lacking the kind of tightly coupled self-regulation that we may admire in individual organisms. Life on Earth is a risky game, then, but with so many players around, there have always been sufficient winners to keep it going, so far. Some puzzlingly 'benign' aspects nevertheless still stand out, such as the relative stability of mean global surface temperatures. It is still too soon to say whether we should put that down to luck, or to some, as yet poorly understood, system of stabilization.

Objectives

When you have finished this book, you should be able to display your understanding of the terms printed in **bold** type within the text as well as the following topics, concepts and principles and, where appropriate, perform simple calculations related to them:

1   the distinction between eukaryotes and prokaryotes, with reference both to structure and to evolutionary scope;

2   evolution by natural selection, including the influence upon it of sexual reproduction, and the consequences of the adaptations to which it gives rise;

3   the biological and paleontological evidence for the main events in the Proterozoic and Phanerozoic evolutionary history of the eukaryotes;

4   the geological and geochemical evidence for the relationship between physical and chemical changes on the Earth and the evolution of life through the Proterozoic and Phanerozoic, especially with reference to the record of isotopic ratios of carbon and strontium preserved in marine limestones;

5   the identification of mass extinctions and major evolutionary radiations from the fossil record, and their possible causes;

6   the links between the evolution of land vegetation and changes in global atmospheric conditions and climate, and the feedbacks involved;

7   use of a simplified planetary energy balance equation to calculate mean global temperatures for given changes in albedo, and the role in this respect of changing land vegetation;

8   the principles involved in constructing qualitative and quantitative models of climate for given paleogeographical arrangements;

9   analysis of the likely circumstances and feedbacks that maintained global icehouse conditions in the late Carboniferous/early Permian, and those associated with the subsequent warming in the late Permian and the mass extinction that came at the end of that period;

10   analysis of the likely circumstances and feedbacks that maintained global greenhouse conditions in the Cretaceous, and the consequences for land vegetation and marine life;

11   the use of fossil land plants in the qualitative and quantitative reconstruction of past climates;

12   the different possible ways in which the relationship between evolving life and the Earth can be interpreted, and approaches to testing between these hypotheses.

References

Ager, D. V. (1993) *The Nature of the Stratigraphical Record* (3rd edn) John Wiley & Sons, Chichester.

Conway Morris, S. (1993) The fossil record and the early evolution of the Metazoa, *Nature*, **361**, 219–25.

Darwin, C. R. (1859) *On the Origin of Species by Means of Natural Selection or the Preservation of Favoured Races in the Struggle for Life*. John Murray, London.

Grotzinger, J. P., Bowring, S. A., Saylor, B. Z. and Kaufman, A. J. (1995) Biostratigraphic and Geochronologic Constraints on Early Animal Evolution, *Science*, **270**, 598–604.

Lovelock, J. E. (1989) Geophysiology, the Science of Gaia, *Review of Geophysics*, **27**(2), 215–22.

Runnegar, B. (1995) Vendobionta or Metazoa? Developments in understanding the Ediacara 'fauna', *Neues Jahrbuch für Geologie und Paläontologie*, **195**, 303–18.

Wolfe, J. A. (1979) 'Temperature Parameters of Humid to Mesic Forests of Eastern Asia and Relation to Forests of Other Regions of the Northern Hemisphere and Australasia.' *United States Geological Survey Professional Paper*, **1106**, 37 pp.

Answers to Questions

Question 1.1

(a) The lunar cratering record suggests that 4000 Ma ago the Earth was experiencing a storm of impacts, which may have repeatedly boiled off any incipient oceans (*Origins of Earth and Life*). You would be ill-advised to leave behind your thermally insulated capsule on this trip, and indeed your time-travel agent might well ask you if you had some alternative destinations in mind. If you risked a visit and succeeded in spotting any standing water through the window of your capsule you might find it fringed with greenish slime. (b) By contrast, you could probably alight from your capsule on the trip to 2000 Ma ago, but not without the breathing apparatus, because of the scarce molecular oxygen, and lashings of suncream, because of the fierce ultraviolet radiation let in by the lack of an effective ozone screen (*Atmosphere, Earth and Life*). Your sandwiches would be preferable to the slimy greenish-grey mats you might encounter along the coast, and the book could be handy unless you really like looking at barren landscapes. (c) Breathing might be manageable, if somewhat strenuous, on the trip to 500 Ma ago, though you could leave the suncream behind now. Unless you took your sandwiches, you would need the crabbing net to catch your lunch at the seaside (but with no guarantees as to what it would taste like). Meanwhile the book would still be useful, as you would have to wait another 100 Ma or so for the appearance of complex and visually interesting life-forms on land (*Atmosphere, Earth and Life*). (d) Top of the list for the trip to 100 Ma ago, of course, would be your weapons, to fend off dinosaurs, if not to help you catch your lunch. (e) The weapons would again prove useful for your trip to 100 000 years ago, the threat now being from large mammals, or even aggressive earlier members of your own species, resenting your presence.

Question 1.2

The genetic material (DNA) in a eukaryote cell is contained in a nucleus, whereas the DNA in a prokaryote cell is confined to a single loop within the cell.

Question 1.3

Single mutations whose effects were hidden by an unaffected partner would still accumulate through the generations, increasing the eventual probability of double mutations at matching sites on the paired chromosomes. To return to the aeroplane analogy, this is rather like the risk of continuing to make further flights in a twin-engined 'plane with only one functioning engine.

Question 1.4

No, because, as discussed above, the focus of natural selection is on the individual genetic entity, who is thus the only *necessary* beneficiary of any adaptations. What happens above that level as a consequence of adaptations of individuals is mere effect, which may be either beneficial or deleterious to the maintenance of the system under consideration.

Question 1.5

If you cast your mind back to *Atmosphere, Earth and Life*, you will recall that it was the long-term burial of organic material that allowed the build-up of molecular oxygen in the atmosphere. Had all the material produced through photosynthesis remained available for respiration (or combustion), it would have consumed the same amount of oxygen as had originally been released in its production. With the removal of such organic material, however, photosynthesis could yield a net excess of oxygen.

Question 1.6
The only likely contenders, all discussed in Section 1.3.1, are (1) cell size (eukaryotes tend to be larger than prokaryotes), (2) presence of organelles, and (3) cellular organization (differentiated cells in multicellular forms). Other, molecular (e.g. genetic), attributes are unlikely to be registered.

Question 1.7
Remember from Section 1.3.4 that the key to understanding the effects of natural selection is the way in which genetic information is transmitted. The coevolved symbionts can be regarded as having started to evolve as an entity from the time when their genetic identities became merged, i.e. when some of the DNA of the prokaryote precursors became incorporated with that in the nucleus of the host.

Question 1.8
The capacity to engulf other cells – the first step towards endosymbiosis – would have necessitated prior loss of the constraining rigid prokaryotic cell wall.

Question 1.9
The episode, in the early Proterozoic, when the deposition of banded iron formations began to decline and red beds started to appear in the rock record – around 2000 Ma ago – is a likely time: these trends in the sedimentary record are interpreted as reflecting the appearance of molecular oxygen in the surface waters of the oceans (*Atmosphere, Earth and Life*).

Question 1.10
The observation implies that sexual reproduction evolved only after those earlier asexual forms had split off. It must have arisen in the common ancestor of the remaining eukaryote groups. The increase in evolutionary flexibility and scope conferred by sexual reproduction was discussed earlier, in Section 1.3.3. What some have referred to as the 'big bang' of eukaryote evolutionary history may, therefore, have been triggered by the appearance of sexual reproduction.

Question 1.11
It was noted in Section 1.4.2 that the grain size of the enclosing sediment determines the resolution possible in preservation of this kind. Thus we could expect no detail finer than the size of sand or silt grains to be preserved, which rules out any prospect of resolving details at or below the cellular level.

Question 1.12
The molecular clock theory supposes a constant average rate of divergence between corresponding sequences. By extrapolating a straight line in Figure 1.15 to the maximum divergence value known among animals (~190%), the time of that original divergence may be estimated (Figure A1). Such an extrapolation yields a date of around 1000 Ma.

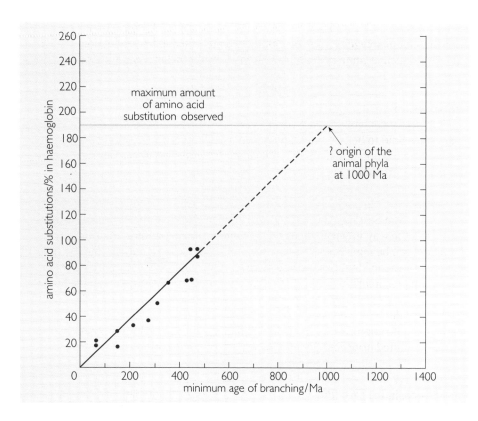

Figure A1
Answer to Question 1.12.

Question 2.1
The various continent–continent collisions would have generated several major mountain chains. Hence one could hypothesize, as has been postulated for the uplift of the Himalayas (*The Dynamic Earth*), that increased weathering of silicate rocks might have drawn down CO_2 from the atmosphere and so led to episodes of climatic cooling.

Question 2.2
For values of $\delta^{13}C_{carb}$ to be positive, there would have to have been relative depletion of the lighter isotope, ^{12}C. Preferential sequestering of ^{12}C is characteristic of photosynthesis (*Origins of Earth and Life*), so the most likely explanation for the overall positive values in the limestones is that large quantities of photosynthesized ^{12}C-enriched organic material were somehow being removed from the marine reservoir. The most likely means for the sustained removal of such material is through sedimentary burial (Section 1.4.1).

Question 2.3
Either the postulated increase in continental weathering (associated with higher values of the strontium isotopic ratio) did not occur, or, if it did, its effects were swamped by those of the increased hydrothermal emissions.

Question 2.4
Banded iron formations (BIFs) are characteristic deposits of the Archean and early Proterozoic (i.e. older than 2000 Ma). They are interpreted as reflecting the free availability of Fe^{2+} in the oceans, and hence pervasive anoxic conditions.

Question 2.5

No. A marked increase in the $^{87}Sr/^{86}Sr$ ratio in the early Vendian suggests a shift in the balance of the origin of strontium in the seawater from the mantle to continental weathering. That this might have been associated with a decrease in the hydrothermal emissions is supported by the apparent absence of banded iron formations associated with the Varanger Ice Age (in contrast to the earlier glaciations).

Question 2.6

Photosynthesis yields molecular oxygen, but any respiratory 'burning up' of the organic material so produced uses it up again. However, if such organic material is removed, without being oxidized, then a net accumulation of molecular oxygen ensues. Thus, the massive burial of organic material that has been postulated for much of the late Proterozoic should have yielded a correspondingly large surplus of molecular oxygen.

Question 3.1

Sharp leaps and falls in $\delta^{13}C_{carb}$ values in marine limestones are shown for this interval, culminating in relatively lower levels in the early Cambrian than in the late Proterozoic. This pattern implies marked fluctuations, followed by an eventual decrease, in the preferential withdrawal of ^{12}C from ocean water, and thus in the rate of burial of organic material.

Question 3.2

Figure 2.1 shows Siberia to have been situated close to the Equator in late Proterozoic times. In the absence of widespread glacial diamictites of late Vendian age, a warm water origin for these low-latitude stromatolites is more likely.

Question 3.3

The differences are quite striking in all respects. (a) Whereas the Ediacaran animals mostly appear to have had rather broad, flat, pancake-like shapes, with either radial ('jellyfish-type') or bilateral ('flatworm-type') symmetry, the mobile forms, at least, in the Burgess Shale were predominantly of elongate or plump, tubular, body form, with bilateral symmetry (looking essentially worm-, or shrimp-like). (b) The latter, moreover, show a fair degree of differentiation of body parts, with their head and tail ends readily distinguishable, and with marked variations in structure along the body. By contrast, the Ediacaran animals show little visible differentiation, consisting either of rather uniform segments along the body, or of simple ridges and grooves arranged radially, spirally or concentrically around a central axis. (c) Many of the Burgess Shale animals also have clear limbs or other projections from the body, but no evidence for limbs can be seen in the Ediacaran forms.

Question 3.4

This animal is likely to have been a predator that could ingest whole animals as prey, because a row of small, similarly shaped conical shells can be seen preserved in its gut. You can see it reconstructed thus as a predator in the right foreground of Figure 3.2.

Question 3.5

By the time of the deposition of the Burgess Shale, a full range of feeding modes had evidently evolved, yielding trophic pyramids with successive layers of consumers, as in modern ecosystems (Figure A2a). Such a range seems to have been lacking among Ediacaran faunas (Figure A2b). There is no evidence either for

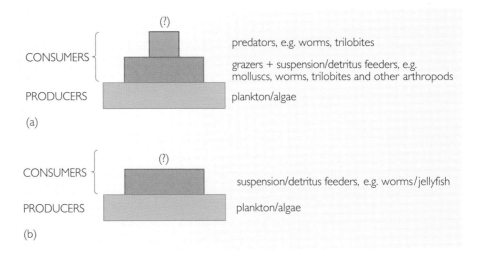

Figure A2
For the answer to Question 3.5. Simplified trophic pyramids for (a) the Burgess Shale fauna, and (b) Ediacaran faunas.

grasping appendages nor for skeletal hard parts with which to tackle large prey. There is nothing to suggest that they were eating each other, at least. Moreover, it has been argued that the commonly intact state of the Ediacaran fossils suggests an absence of large-scale predators, scavengers and deep burrowers.

Question 3.6
The small shelly fossils of earliest Cambrian age include the shells of tiny molluscs (Figure 3.6a), only a fraction of a millimetre across.

Question 3.7
At present, none of the scenarios can be categorically rejected. The Namibian finds (Section 3.2.2) show that Ediacaran faunas persisted through to the end of the Vendian, and a few types, moreover, even survived into the Cambrian (Section 1.5.1). Hence the possibilities remain for either an evolutionary connection, or direct ecological interaction entailing displacement, between the two sets of animals. In either case, however, the transition would have had to have been rapid, on a geological time-scale. Alternatively, a mass extinction of most of the Ediacaran animals at the close of the Vendian could have been followed by the rapid radiation of Cambrian animals, derived from unknown ancestors.

Question 3.8
Anoxia in deeper waters, brought about by a variety of means, was the main factor that favoured the burial of organic matter. It prevented aerobic decomposers from breaking down the remains of dead plankton, as they sank to the sea floor from oxygenated surface waters (Section 2.3.2). Rising levels of molecular oxygen in the Phanerozoic could be expected to have diminished the occurrence and the extent of oceanic anoxia.

Question 3.9
The churning of the sediment (which would have destroyed faecal pellets), and the associated recycling of its carbon content into the overlying water, would have counteracted the hypothesized bypassing effect of faecal pellets.

Question 3.10
This peak represents the families of soft-bodied animals, or of those which possessed only thin unmineralized skeletons, exceptionally preserved in the Burgess Shale.

Question 3.11

Following on from that in the late Ordovician (1), comparable sharp falls in diversity occurred in (2) the late Devonian, (3) the late Permian, (4) the late Triassic and (5) the late Cretaceous. It might seem at first glance that the late Triassic extinction was not really on a par with the others. However, the *relative* drop in diversity then was still considerable, given the depleted numbers of families that remained after the particularly severe late Permian extinction. (These are discussed in further detail in Box 3.2.)

Question 3.12

After each mass extinction, diversity rebounded again relatively rapidly. Such a consistent response accords well with the hypothesis discussed in connection with the Cambrian explosion – that of unrestrained evolutionary radiations filling vacated habitats (Section 3.2.3).

Question 3.13

Following the trends proposed, rates of origination (declining) and extinction (increasing) could eventually come to match one another, as diversity rose. Thereafter, because of the balanced budget of gains and losses, diversity would then stay at the same level (or at least hover around it, if random fluctuations were allowed for).

Question 3.14

No. Each successive fauna appears to have been proportionately less drastically affected by mass extinction than its predecessor. This is most clearly seen in the case of the great mass extinction at the close of the Permian. Here, family diversity in the Paleozoic Fauna can be seen to have plummeted (the Cambrian Fauna being already of negligible diversity at this time), while the Modern Fauna was much less profoundly affected. You can check this on the separate curves shown in Figure 3.11. Again, the earlier Paleozoic (late Ordovician and late Devonian) extinctions, for example, all cut relatively more deeply into the diversity of the Paleozoic Fauna than into that of the Modern Fauna. Moreover, the late Ordovician extinction almost halved the diversity of the Cambrian Fauna, compared with the loss of just under a third of the families in the Paleozoic Fauna.

Question 3.15

As they sank, on death, to the sea floor, they would have augmented carbonate sedimentation yet further offshore, and in deeper water, continuing the trend noted earlier for the Cambrian radiations.

Question 4.1

Growing mostly on soils that overlie weathering rock or sediment, land vegetation, which often has deep roots, is advantageously placed to sequester (and recycle) the nutrients released by the weathering. By contrast, much of the open ocean is virtually a 'nutrient desert'. Moreover, light penetrates only the surface layers of the ocean (the photic zone), so plankton growth is restricted to the surface waters, and relies heavily on localized nutrient-bearing upwellings, as you saw in *The Dynamic Earth*.

Question 4.2

Most freshwater aquatic plants are secondarily re-adapted to an aqueous environment. They are not directly descended from previously ocean-dwelling plants but have ancestors that first invaded the land and subsequently moved back into the water. They retain many of their former adaptations to a dry-land existence.

Question 4.3

The reverse process is *respiration*. Both plants and animals use respiration to oxidize carbohydrates to produce carbon dioxide and to release energy which powers life processes. Thus:

$$C_6H_{12}O_6 + 6O_2 \longrightarrow 6CO_2 + 6H_2O + energy$$

Question 4.4

It is an example of *homeostasis*, which was discussed in Section 1.3.4.

Question 4.5

It can be supposed from Figure 4.14 that sufficient CO_2 to meet photosynthetic requirements (limited by other factors) could have entered the plants through relatively few stomata. If water was the limiting factor for photosynthesis, then having only a few stomata would have been advantageous because it would have reduced water loss to the atmosphere.

Question 4.6

First, as you have seen earlier in the Course, the further back we go in time, the weaker the Sun would have been (the Faint Young Sun paradox), so the value for solar energy flux used in Equation 4.1 would be reduced by a few per cent. Secondly, for simplicity, we have assumed there were no clouds; but evaporation from the oceans would have produced some clouds, particularly over the otherwise quite dark ocean surface, even though, as we have argued, the lack of vegetation would have reduced the amount of water vapour in the atmosphere. The effect of clouds on the global energy budget is complex, but they would have increased the albedo by a significant amount.

Both these factors would reduce the amount of radiation reaching the surface of the Earth, and so would lead to a lower mean global temperature.

Question 4.7

From Equation 4.1 as before,

$$T = \frac{\dfrac{1365}{4} \times (1 - 0.0733) - 210}{2.1} \, °C = 50.6°C$$

So, your calculations should have given the value of 50.6 °C.

Question 5.1

These pressure belts are not the direct result of the thermal gradient but of the dynamics of air circulation on a rotating sphere. Because of the rotation of the Earth, air rising at the Equator and flowing towards the poles is deflected eastward (the Coriolis effect) while air sinking at the poles and flowing equatorward is deflected westward. By about 30° latitude the mean flow of air is parallel to the Equator and further poleward progression is clearly impossible. Heat exchange between the Equator and the poles is effected by an intermediate cell approximately between 30–35° and 55–60°. Where the edge of equatorward-moving cold air (the polar front) meets the poleward moving air at 55–60° the air is forced to rise, so forming a low-pressure zone and belt of precipitation.

Question 5.2

Areas of low pressure are the result of warm air rising. Because the air is warm it can hold more moisture than cool air. However, as it rises, it also experiences less confining pressure, so it expands and cools. The water vapour in the air then condenses, clouds form and the droplets coalesce to form rain.

Question 6.1

In contrast to today, Figure 6.1 shows that there was one enormous supercontinent, Pangea (*The Dynamic Earth*), which incorporated all but some Asian microcontinents and which extended from the South Pole to high northern latitudes.

Question 6.2

The two main reactions considered involve the weathering of calcium carbonate ($CaCO_3$) and silicate ($CaSiO_3$) rocks. The first takes in one molecule of CO_2 for each molecule of $CaCO_3$ weathered, but, because the ensuing precipitation of carbonate releases it again, there is no net drawdown of CO_2:

$$CaCO_3 + CO_2 + H_2O \underset{\substack{\text{carbonate} \\ \text{precipitation}}}{\overset{\substack{\text{carbonate} \\ \text{dissolution}}}{\rightleftarrows}} Ca^{2+} + 2HCO_3^{-}$$

The second, silicate weathering, involves two molecules of CO_2 for every silicate molecule weathered:

$$\underbrace{CaSiO_3}_{\substack{\text{silicate} \\ \text{rock}}} + \underbrace{2CO_2 + 3H_2O}_{\text{from rainwater}} \longrightarrow \underbrace{Ca^{2+} + 2HCO_3^{-} + H_4SiO_4}_{\text{stream/river water}}$$

But since the ensuing precipitation of carbonate releases one molecule of CO_2 (see the carbonate precipitation reaction above), there is therefore a net drawdown of CO_2 from the atmosphere when silicate rocks are weathered.

Question 6.3

The presence of a continental mass over the South Pole is similar to the model of the 'cap world'.

Question 6.4

Such mountain building would have resulted in an increase in the amount of rock available to be eroded. This, in turn, would have led to more CO_2 being sequestered from the atmosphere as rock silicates were weathered, ultimately ending up in the oceans, as $CaCO_3$.

Question 6.5

Increased weathering of continental rocks occurred due both to the effects of vegetation and prominent mountain belts. Also there was probably less hydrothermal activity in the superocean (as there were fewer mid-ocean ridges). Therefore $^{87}Sr/^{86}Sr$ ratios should have been relatively high.

Question 6.6

There was one superocean in the Permo-Carboniferous interval with a consequent reduction in the total length of the mid-ocean ridge system. Remember also that $^{87}Sr/^{86}Sr$ ratios were relatively high at this time, reflecting the lower sea-floor spreading rates. These factors suggest that global sea-level would have been relatively low.

Question 6.7

Remember, as shown in from Figure 6.2, that earlier Paleozoic levels of atmospheric CO_2 were probably much higher, resulting in a substantial greenhouse effect which could have swamped the cooling effect of a lower solar flux.

Question 6.8
Think back to the monsoonal system generated by the formation and movement of Pangea. You saw in Section 6.3.3 that more arid climates gradually spread eastwards – a trend reflected in the rock record. The same thing affected vegetation, so that plants adapted to dry conditions dominated first in Europe and North America, but much later further eastwards in places such as China. From Figure 6.1b, you can see that another important factor would have been the relatively small size of the eastern low-latitude landmasses – islands surrounded by an ocean which would have produced a constant source of moist air.

Question 6.9
Recall from Section 3.4.4 that members of the Paleozoic Fauna (dominated by exposed sessile bottom-dwelling forms) suffered badly at the end of the Permian and never re-attained their former diversity. Members of the Modern Fauna (with more free-living and burrowing forms, including numerous predators specializing on shelly prey) also suffered, but successfully rediversified afterwards.

Question 6.10
A spectacular accumulation, the Siberian flood basalts, was discussed in *The Dynamic Earth*. They constitute one of the largest continental examples of the Phanerozoic, and were erupted near to the Permian–Triassic boundary.

Question 6.11
In contrast to the Permo-Carboniferous icehouse phase, when the burial of vegetation at low latitudes was sequestering CO_2 on a massive scale, the later Permian exposure and weathering of this store (i.e. the reverse process), would have released (isotopically light) CO_2 back to the atmosphere.

Question 6.12
The large-scale oxidation of organic matter exposed to erosion could have reduced atmospheric oxygen levels.

Question 7.1
First, the rifting and break-up of Pangea would have created several new mid-oceanic ridge systems by Cretaceous times, the combined volume of which would have displaced corresponding amounts of water from the ocean basins, as explained in *The Dynamic Earth*. Second, the water that in late Carboniferous to early Permian times was tied up in the South Polar continental ice sheet would, in the Cretaceous, have been back in the oceans.

Question 7.2
In relatively warm, dark situations, such as those of the Cretaceous, food use through respiration would have been greater than food production by photosynthesis, and evergreen plants would have starved to death.

Question 7.3
The large number of uniform earlywood cells shows that the tree experienced little variation in growth conditions throughout the growing season. If we assume they were produced over a six-month (light) period, then a new cell was formed every few days. By contrast, only a few latewood cells were produced, which implies that the period of growth reduction prior to dormancy was very short. Figure 7.3 shows that at high paleolatitudes (>75°) the transition between constant daylight and constant darkness takes place over a period of just a few weeks. If latewood was

produced during that transition (or began several weeks before, as the Sun got lower and lower in the sky), then the small amount of latewood suggests a similarly rapid transition in the Cretaceous.

Question 7.4
The rings of the latest Cretaceous specimen look narrower, i.e. there are fewer earlywood cells in each ring. However, if you look at the latewood cells you will also see a difference in their appearance. In some rings, for example the upper two, the latewood cells become progressively thicker-walled up to the ring boundary; thereafter the following earlywood cells show an abrupt change to thin walls and large cavities again. Others, for example the rings labelled 2 and 4 in Figure 7.10, are characterized by only a few latewood cells that are not developed gradually, but instead appear as distinct bands in what would otherwise be earlywood.

Question 7.5
One possible explanation for less benign summers would be that Alaska drifted north during the late Cretaceous. According to some researchers this indeed did happen. Another possible explanation for the decline in summer temperatures, however, might be a general global cooling. Again, there is evidence for this from elsewhere in the world. Therefore, both processes seem to have been operating.

Question 7.6
The following adaptations are displayed in *Pseudofrenelopsis*:
1   a thick cuticle;
2   very small leaves resulting in a low surface area to volume ratio for the plant as a whole;
3   stomata sunken in deep pits, so capturing still moist air close to the stomatal apparatus and guard cells;
4   finger-like projections covering the stomatal pit that could close off the pits in extreme drought;
5   hairs along segment margins that could have aided extraction of water from moist air.

Question 7.7
The percentage of entire margined leaf species for this time was $(22/67) \times 100\% = 32.8\%$. Therefore the northern Alaskan MAT at this time would have been 10 °C.

Question 7.8
From the available information, the MAT could fall anywhere between 3 °C and 13 °C, and the MART between 0 °C and 35 °C. The forest closely resembles a mixed coniferous forest. In fact, the modern forest type most closely related to that of the mid-Cretaceous of northern Alaska is a mixed coniferous forest which grows at low altitudes in mountainous regions and yields a MAT of 9–11 °C and a MART of 15–22 °C.

Question 7.9
The subdued relief, together with the aridity if the hinterland would have meant that supplies of land-derived sands and muds, carried in by streams and rivers, would have been limited, and associated largely with the localized zones of uplift. By contrast, the breakdown of calcareous shells produced by the myriads of organisms thriving in the broad expanses of warm shallow sea (the 'shallow water carbonate factory' mentioned in *The Dynamic Earth*) would have generated copious amounts of carbonate sediments, eventually forming limestone.

Question 7.10

Extensive glaciations, over the last couple of million years, have caused several global ('glacio-eustatic') falls in sea-level on the scale of a hundred metres or more (*The Dynamic Earth*). Although there were probably small mountain glaciers in Cretaceous high latitudes, forests were spread over lowland areas, instead of ice, as you saw in Section 7.3.

Question 7.11

The volume of platform limestone being buried per million years would have been about

$$9.68 \times 10^6 \, \text{km}^2 \, \text{(area)} \times 0.03 \, \text{km (thickness)} = 0.29 \times 10^6 \, \text{km}^3$$

This is equivalent to a rate of burial of

$$(0.29 \times 10^6 \times 10^9) \, \text{m}^3 \times (2.71 \times 10^3) \, \text{kg m}^{-3} = 0.79 \times 10^{18} \, \text{kg of limestone per million years, or } 0.79 \times 10^{12} \, \text{kg yr}^{-1}$$

The rate of burial of carbon in this would thus have been

$$0.12 \times (0.79 \times 10^{12}) \, \text{kgC yr}^{-1} = 0.09 \times 10^{12} \, \text{kgC yr}^{-1}$$

Question 7.12

With such a substantial rise in sea-level, it is likely that the calcium carbonate compensation depth also rose. Hence rather less ocean floor would have accumulated deposits of calcium carbonate than is the case today. Drilling of the ocean floor nevertheless confirms that plenty was deposited.

Question 7.13

The relationship between organic carbon burial and $\delta^{13}C_{\text{carb}}$ values was discussed in Chapter 2. Since the buried organic material tends to be relatively enriched in the lighter isotope, ^{12}C, episodes of such enhanced burial would have led to increased seawater values of $\delta^{13}C$, in turn reflected in marine sediments such as limestones.

Question 7.14

Only the burial of organic carbon involves a net supply of molecular oxygen to the atmosphere; the burial of carbonate carbon would have had no such effect, involving only the incorporation of molecular CO_2 into the carbonate (CO_3^{2-}) ions of the carbonate rocks. On the other hand, the episodes of enhanced organic carbon burial noted earlier would have increased the supply of molecular oxygen. What became of this is uncertain at present, though you may have already seen in *Atmosphere, Earth and Life* that Berner and Canfield's best estimates of atmospheric oxygen levels recognize a distinct rise through the Cretaceous.

Question 7.15

The relative eustatic flooding of the continents, together with the modest degree of mountain building, discussed at the start of this chapter, make any increased weathering of fossil fuels an unlikely explanation. But the resulting decreased area of exposed silicates would have at least reduced one alternative sink for CO_2, although the warmer temperatures would have counteracted that to some extent. However, as shown in *The Dynamic Earth*, the total estimated loss to that sink today is only $0.03 \times 10^{12} \, \text{kgC yr}^{-1}$. Even if that were reduced to zero, it would still be inadequate to account for the extra carbon burial in the Cretaceous.

Question 7.16

The carbon outflow through the platforms would have temporarily failed to match the increased flux of CO_2 into the atmosphere. Unless other outflows compensated, there would then have been an increase in CO_2 level in the atmosphere.

Question 8.1

This Course has provided you with numerous possible answers. Foremost among these is atmospheric composition, and in particular the abundance of molecular oxygen (*Atmosphere, Earth and Life*), though the suppression of carbon dioxide levels has been no less important, especially in relation to climate (Chapters 6 and 7). You may also have thought of major chemical changes in the sea and on the land – the oceanward shift in carbonate deposition (Chapters 3 and 7) and the massive acceleration of chemical weathering processes (Chapter 6), for example. Physical influences include the effects on marine sediments of burrowing (Chapter 3), or of platform or reef-building organisms (Chapter 7) and, on land, the climatic effects of plant transpiration (Chapter 4). And so the list could go on.

Question 8.2

There is no theoretical justification for such an expectation. Recall from Section 1.3.4 that the focus of natural selection is on the individual genetic entity, who is thus the only *necessary* beneficiary of any adaptations. Anything beyond that is an incidental effect. Since the Earth and its life as a whole neither dies nor reproduces as individuals do, it cannot have evolved by means of natural selection between varied genetic entities. Hence, we need not expect any integrated co-adaptation of its parts such as is found in living organisms.

Question 8.3

The case study in Chapter 6 shows that the consequences of feedbacks were quite variable, and included some distinctly destabilizing positive feedbacks. The transition from the Permo-Carboniferous icehouse world to the warmer world of the late Permian, for example, seems, if anything, to have been helped on its way by the weathering of some of the earlier-formed coals, releasing CO_2 (Section 6.6.3), while 'the loss of most coal-swamp vegetation may have provided another positive feedback to the warming, since there would have been less capacity for CO_2 removal from the atmosphere' (Section 6.5). In this case, it could hardly be said that the new regime was being 'regulated at a state tolerable for the biota', as it eventually culminated in the largest mass extinction of the Phanerozoic (Section 6.6).

Question 8.4

Recall from Section 1.4.2 that the eukaryote ancestors are thought to have been anaerobes, which formed a symbiotic union with aerobically respiring prokaryotes, as an adaptive response to the rise in the level of atmospheric molecular oxygen. So successful was this union in oxidizing conditions that the endosymbiotic prokaryotes became fully integrated as mitochondria (so various lines of evidence suggest), and subsequently nearly all eukaryotes have been aerobic.

Acknowledgements

We are most grateful for the particularly detailed and constructive comments on earlier drafts provided by the external assessor for the book, Professor Simon Conway Morris, and by Dr Steve Drury (internal course reader). We also wish to thank the following for their helpful comments and advice: Professor Bill Chaloner (Course Assessor), Fran Van Wyk de Vries, Peter Daniels, Margaret Deller, Jim Grundy, Colin Whitmore (student readers), Cynthia Burek (tutor reader), as well as other members of the Course Team. The authors and book chair must remain responsible, however, for any errors that remain. Finally, a special thank you to John Watson, of the Open University, who produced the lively reconstructions (Figures 3.2, 3.12, 6.5a, 7.6 and 7.15).

Grateful acknowledgement is made to the following sources for permission to reproduce material in this book:

Figures

Cover photograph, frontispiece, Figures 6.6c, 7.19, 7.23: Peter Skelton, Open University; *Figures 1.1a, 1.9:* Andrew Knoll, Harvard University; *Figures 1.1b, 1.14c,d:* Peter Crimes, Liverpool University; *Figure 1.4:* Biophoto Associates; *Figure 1.10:* Butterfield, N. J. and Chandler, F. W. (1992) 'Palaeoenvironmental distribution of proterozoic microfossils, with an example from the Agu Bay formation, Baffin Island', *Palaeontology,* **35**, Part 4, November 1992, © The Palaeontological Association; *Figures 1.14a,b, 3.3, 3.6b,c:* Simon Conway Morris, University of Cambridge; *Figures 1.15, A1:* B. Runnegar © Norwegian University Press; *Figure 2.1:* Dalziel, I. W. D. (1995) 'Earth before Pangea', *Scientific American,* January 1995. Scientific American, Inc. All rights reserved. Illustrations by Ian Worpole; *Figure 2.2a:* Michael Hambrey, Liverpool John Moores University; *Figure 2.2b:* Cynthia Burek; *Figure 2.2c:* Ian Fairchild, Keele University; *Figure 2.3:* adapted from Knoll, A. H. (1994) in Bengtson, S. (ed.) *Early Life on Earth – Nobel Symposium No. 84.* Copyright © Columbia University Press. Reprinted with permission of the publisher; *Figure 3.1:* Reprinted with permission from Bengtson, S. and Yue Zhao 'Predatorial borings in Late Precambrian mineralized exoskeletons', *Science,* **257**, pp 367–369. Copyright 1992 American Association for the Advancement of Science; *Figure 3.4a:* Prof H. B. Whittington, University of Cambridge, (from the collection of the Geological Survey of Canada); *Figure 3.5a:* The Natural History Museum, London; *Figure 3.5b:* Cisne, J. L. (1975) *Fossils and Strata,* Vol 4, Universitets Forlaget, Oslo; *Figure 3.7:* Reprinted with permission from *Nature,* **361**, p. 220, Simon Conway Morris 'The fossil record and the early evolution of the Metazoa', Copyright 1993 Macmillan Magazines Ltd; *Figures 3.8, 3.11:* Sepkoski, jr, J. J. (1990) 'Evolutionary faunas', in Briggs, D. E. G. and Crowther, P. R. (eds) *Palaeobiology A Synthesis,* Blackwell Science Ltd. Adapted from Sepkoski, jr, J. J. (1984) 'A kinetic model of Phanerozoic taxonomic diversity. III. Post-Paleozoic families and mass extinctions', *Paleobiology,* **10**, pp 246–267; *Figure 3.10:* Reprinted with permission from Benton, M. J. 'Diversification and extinction in the history of life', *Science,* **268**, p. 53. Copyright 1995 American Association for the Advancement of Science; *Figures 4.3, 4.6, 4.11a, 4.13, 4.16, 4.17a, 6.6b,d, 6.8, 7.4, 7.5, 7.7, 7.8, 7.9, 7.10, 7.11b, 7.12:* Bob Spicer, Open University; *Figure 4.4c,d:* Prof W. G. Chaloner, FRS, Royal Holloway College; *Figure 4.5:* Dr Jeremy Burgess/Science Photo Library; *Figure 4.8:* Stewart, W. N. and Rothwell, G. W. (1993) *Paleobotany and the Evolution of Plants,* 2nd edn, Cambridge University Press, adapted from Andrews, H. N. (1960) 'Notes on Belgium specimens of sporogonites', *Palaeobotanist*, **7**; *Figure 4.9a:* Stewart, W. N. and Rothwell, G. W. (1993) *Paleobotany and the Evolution of Plants*, 2nd edn, Cambridge University Press, adapted from Edwards, D. (1970) 'Fertile rhyniophytina from the Lower Devonian of Britain', *Palaeontology*, **13**; *Figure 4.9b:* Professor Diane Edwards, University of Wales; *Figure 4.10a:* Stewart, W. N. and Rothwell, G. W. (1993) *Paleobotany and the Evolution of Plants*, 2nd edn, Cambridge University Press, adapted from Edwards, D. and Edwards, D. S. (1986) 'A reconsideration of the rhyniophytina banks', in Spicer, R. A. and Thomas, B. A. (eds) *Systematic and Taxonomic Approaches in Palaeobotany, Systematics Association Special Volume*, **31**; *Figure 4.10b:* Stewart, W. N. and Rothwell, G. W. (1993) *Paleobotany and the Evolution of Plants*, 2nd edn, Cambridge University Press; *Figure 4.10c:* Stewart, W. N. and Rothwell, G. W. (1993) *Paleobotany and the Evolution of Plants*, 2nd edn, Cambridge University Press, adapted from Kasper, A. E. & Andrews, H. N. (1972) 'Pertica, a new genus of Devonian plants from northern Maine', *American Journal of Botany*, **59**; *Figure 4.11b:* Stewart, W. N. and Rothwell, G. W. (1993) *Paleobotany and the Evolution of Plants*, 2nd edn, Cambridge University Press, adapted from Ananiev, A. R. and Stepanov, S. A. (1968) 'Finds of sporogenous organs of psilophyton princeps Dawson emend. Halle in the Lower Devonian of the south-minusinsk hollow', *Treatises Tomsk Order Red Banner of Labor State University Geological Series*, **202**; *Figures 4.11c,d:* Stewart, W. N. and Rothwell, G. W. (1993) *Paleobotany and the Evolution of Plants*, 2nd edn, Cambridge University Press, adapted from Doran, J. B. (1980) 'A new species of psilophyton from the Lower Devonian of northern New Brunswick', *Canadian Journal of Botany*, **58**, National Research Council of Canada; *Figure 4.17b:* Beck, C. B. (1971) 'On the anatomy and morphology of lateral branch systems of Archaeopteris', *American Journal of Botany*, **58**, Botanical Society of America, Inc; *Figure 4.17c:* Beck, C. B. (1962) 'Reconstructions of Archaeopteris and further consideration of its phylogenetic position', *American Journal of Botany*, **49**, Botanical Society of America, Inc; *Figures 5.4, 5.5, 7.16, 7.17:* Paul Valdes, University of Reading; *Figure 6.1:* Martini, I. P. (1996) *Late Glacial and Postglacial Environmental Changes – Quaternary Carboniferous – Permian and Proterozoic*, Oxford University Press, Inc. Reprinted by permission of Oxford University Press, Inc; *Figure 6.2:* Berner, R. A. (1994) '3 Geocarb II: A revised model of atmosphere CO_2 over Phanerozoic time', *American Journal of Science*, **294**, January 1994, reprinted by permission of the American Journal of Science; *Figure 6.3:* Raymo, M. E. (1991) 'Geochemical evidence supporting T. C. Chamberlin's theory of glaciation', *Geology*, **19**, (4); *Figure 6.4:* Reprinted from *Tectonophysics*, **222**, Crowley, T. J. 'Climate change on tectonic time scales', pp 277–294, Copyright 1993 with kind permission from Elsevier Science – NL Sara Burgerhartstraat 25, 1055 KV Amsterdam, The Netherlands; *Figure 6.6a:* Phillips, T. L. and DiMichele, W. A. (1992) 'Comparative ecology and life-history biology of arborescent lycopsids in Late Carboniferous swamps of Euroamerica', *Ann. Missouri. Bot. Gard*, **79**; *Figure 6.7:* DiMichele, W. A. and Aronson, R. B. (1992) 'The Pennsylvanian-Permian vegetational transition: A terrestrial analogue to the onshore-offshore hypothesis', *Evolution*, **46**, (3), Allen Press Inc; *Figure 6.9:* Holser, W. T. and Magaritz, M. (1987) 'Events near the Permian–Triassic boundary', *Modern Geology*, **11**, Gordon and Breach Science Publishers; *Figure 6.10:* Holser, W. T. and Schönlaub, H. P. (1991) 'The Permian–Triassic boundary in the Carnic Alps of Austria (Gartnerkofel region)', *Abhandlungen der Geologischen Bundesanstalt*, **45**, Copyright by Verlag der Geologischen Bundsanstalt, Wein, Austria; *Figures 6.12, 6.13:* From *The Great Paleozoic Crisis,* by D. H. Erwin. Copyright © 1993 by Columbia University Press. Reprinted with permission of the publisher; *Figures 7.1, 7.2:* Reprinted from

Palaeogeography, Palaeoclimatology and Palaeoecology, **40**, Parrish, J. T., Ziegler, A. M. and Scotese, C. R. 'Rainfall patterns and the distribution of coals and evaporites in the Mesozoic and Cenozoic', p. 88, Copyright 1982 with kind permission from Elsevier Science – NL Sara Burgerhartstraat 25, 1055 KV Amsterdam, The Netherlands; *Figures 7.3, 7.13, 7.14:* USGS Printing Office, Washington DC; *Figures 7.11a,c,d:* Dr Joan Watson, Manchester University; *Figure 7.18:* Sohl, N. F. (1987) 'Presidential address – Cretaceous gastropods: Contrasts between Tethys and the temperate provinces', *Journal of Paleontology*, **61**, (6), The Paleontological Society; *Figure 7.24:* J. A. Burnett, University College, London; *Figure 7.25:* Jenkyns, H. C., Gale, A. S. and Corfield, R. M. (1994) 'Carbon and oxygen-isotope stratigraphy of the English chalk and Italian scaglia and its palaeoclimatic significance', *Geological Magazine*, **131**, Cambridge University Press.

Index